JOHN WRIGHT is a writer and naturalist who has led over a thousand foraging walks in the last thirty years. His books include *The Forager's Calendar*, a bestseller on the delights and practicalities of foraging in any season; *The Observant Walker*, which highlights the wild food, natural wonders and human histories that can be found on any walk; and *A Natural History of the Hedgerow: and ditches, dykes and dry stone walls*. He lives with his wife in Dorset.

Praise for *The Forager's Calendar*

'John Wright writes as though he's talking directly to you, a good friend in the same room. His harvest of fascinating information is worn lightly, with funny, whimsical observations' *Countryfile Magazine*

'A hugely useful, well-illustrated and often funny book' *The Times*

'He writes so engagingly that it's hard to imagine that actual foraging can be more attractive than reading his accounts of it' John Carey, *Sunday Times*

'John Wright is an authoritative and often funny guide ... by the end he makes finding one's dinner from the earth sound immensely appealing' *Guardian*

Praise for *A Natural History of the Hedgerow*

'This illustrated survey is historically detailed, enriched by the author's deep knowledge of British landscapes and natural history' *Guardian*

'A true labour of love spiced with a fine dry humour ... [not] just a delightful one-off read, but an invaluable work of reference that will remain on my bookshelves for good' Christopher Hart, *Sunday Times*

Also by John Wright

*The Observant Walker: Wild Food, Nature and
Hidden Treasures on the Pathways of Britain*

*The Forager's Calendar: A Seasonal Guide to
Nature's Wild Harvests*

*A Natural History of the Hedgerow: and ditches,
dykes and dry stone walls*

A Spotter's Guide to the Countryside

Uncovering the wonders of Britain's woods, fields and seashores

John Wright

PROFILE BOOKS

This paperback edition first published in 2023

First published in Great Britain in 2021 as *A Spotter's Guide to Countryside Mysteries* by
Profile Books Ltd
29 Cloth Fair
London
EC1A 7JQ

www.profilebooks.co.uk

All images are courtesy of the author unless specified below.

Images on pp. 146, 150 and 198 © Bryan Edwards
p. 34 Stan Pritchard / Alamy Stock Photo; p. 39 Andy
Harmer / Alamy Stock Photo; p. 66 image released into the
public domain by the copyright holder (via Wikipedia);
p. 69 Powered by Light/Alan Spencer / Alamy Stock
Photo; p. 101 Andrew Darrington / Alamy Stock Photo;
p. 197 Tony Watson / Alamy Stock Photo; p. 202 Nancy
Morrison / Alamy Stock Photo

Permission to quote from *Watership Down* by Richard
Adams (Puffin Books) granted by David Higham Associates.

1 3 5 7 9 10 8 6 4 2

Printed and bound in Great Britain by Bell and Bain Ltd,
Glasgow

Design by James Alexander/www.jadedesign.co.uk

FSC
www.fsc.org
MIX
Paper from
responsible sources
FSC® C018072

A CIP catalogue record for this book is
available from the British Library.

ISBN 978 1 78816 8274
eISBN 978 1 78283 862 3

For Lily and Bill

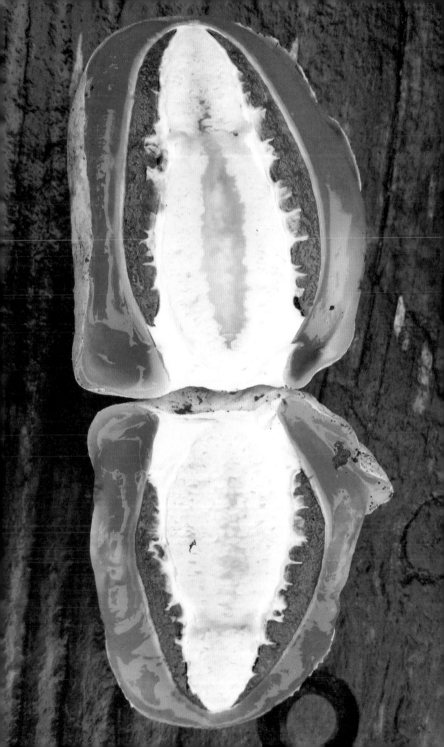

Introduction

There are so many questions that attend a walk in the countryside. Why is this field laid out the way it is? What is a pond doing on the top of that chalk hill? How would this water meadow have looked in its heyday, and how did it operate? Walking through a forest, we might look up and wonder what the mass of twigs in the middle of that tree might be, or the enormous bulge on the trunk of another. As we set out across country, we might ask what are those thousands of grassy bumps in this damp field, and what are those rings of dark grass? And as we sit down for a rest, we may ponder why a traditionally green grass grows a white sheath on its stem, and where on earth did that tiny orange lollipop sticking out of the ground come from? All I can ever do is chip away at these and a thousand other mysteries that have puzzled me over the many years.

Countryside mysteries came to me early. In 1965 I discovered a strange black hemispherical lump growing on a dead ash tree. It looked as though it had been crafted from charcoal; indeed, it possessed the same crumbly texture. This black blob seemed so incongruous, so very dead, that I wondered how it could grow, if grow it ever did. My reaction was not so much of curiosity, though certainly I was curious, but out of *annoyance* that there was something for which I could not even guess an explanation. Determined to address this humiliating failure in understanding, I looked it up in the local library and discovered it to be a fungal fruiting body commonly known as a Cramp Ball, that it seems to grow layer by layer and that it is sometimes covered with a brown velvet. Its whole story is *much* more complex (the 'dead' phase, for example, is the most active), its mystery deeper, as I will explain later. A much earlier fungal conundrum (though one that was a matter of how something familiar behaved, rather than what it was) came courtesy of my grandmother, who insisted that field mushrooms only grow overnight. Even to your then five-year-old author this seemed an unlikely claim, and it worried me for years.

The above-mentioned black fungus is a riddle presented to us by nature. Such things were once understood by no one, their character to be revealed only by diligent investigation. But there are also things in the countryside whose origins and purpose we have simply forgotten: the constructions and shadows of constructions left to us by our ancestors. Mysterious holes in the ground, strange embankments around our coastline, piles of stones in the middle of a field and pathways laboriously and pointlessly dug into a hill. All demand an explanation; all have a history that illuminates the lives of those who were here before us.

I still see things every week, natural or unnatural, for which the question 'What on earth is going on here?' is the first that comes to mind, though not always so politely expressed. It could be a branch that should be smooth but which is deeply furrowed, a blackened centre to a flower where the bright yellow anthers should be, a group of trees arranged in a hollow parallelogram, an aerial photograph of a patchwork landscape where no fields exist, or, of course, another strange lump on a tree.

It is quite possible to dismiss such questions with a name and the class to which the mystery in question belongs. Perhaps a little more detail would be in order, such as how it acquired that name or what its relationship with man might be. That the blackened centre of the flower, for example, is caused by a smut, that it is so called because it looks like soot, and it is a type of fungus that causes problems for farmers. Few, however, would or should tolerate such short change, demanding instead a much deeper understanding; they want the whole story. A visitor from the seventeenth century, confronted by a Ford Ranger pickup (to take an example close to my heart), would no doubt be fascinated to hear that there are tens of thousands of them, that they can carry a ton of bricks, that they come in grey, blue and white and you look really cool when driving one. But he or she would not be satisfied with such superficial stories; the question that cries out for an answer is: 'What makes the damnable thing move?'

All living organisms have a complexity beyond the scope of any book. With my natural mysteries, explanations below a certain depth are superfluous. I may write of fungal spores being produced on a fungal fruiting body, but I make no attempt to describe *how* they are produced. In general, the technical explanations that I provide in this book address the aspect of the organism that makes them mysterious in the first place. Several of these explanations address the close relationships between organisms: symbioses. These are often intricate balancing acts and a joy to contemplate. The white sheath seen on grass stems, for example, is part of an 'arrangement' between a grass, a fungus, a fly and a bacterium.

With man-made mysteries it may be the details of how something works, when and why it was used and why it faded into history. Many know what water meadows look like and perhaps that they involve the early flooding of a meadow for early grazing. But there is much more to them than this: the complex details of how they operated and, more important, the fact that they were once part of a larger system of agriculture for which the primary concern was not grazing animals but the growing of corn.

Most things were a mystery at some point in history, so I have limited my topics to those that are interesting in themselves and for which most people would be bereft of an explanation for what they are or how they work. The dire alternative (one beyond my ability to write and your patience to read) is a complete treatise on British biology, geology, archaeology and agricultural history. My further self-imposed restriction in describing man-made objects is to address, for the most part, those things with an agricultural inclination. Ancient monuments are well recorded and are well described in the British literature. There are no barrow mounds or temples here. I have also left out most historic mining operations, restricting myself to the agriculturally inclined marl pit.

Many things have, therefore, not found a place in this book, and you have my apologies if I have not explained something that has always puzzled you, such as what the 'groynes' one finds on beaches are for, or (the big one) what is the point of wasps? Those subjects that I have included are a very personal choice in a very personal book, so if you cannot find an answer here, I encourage you to investigate yourself – it is great fun.

To 'spot' something, you must be in the type of location it favours. I have therefore arranged my subjects in three very broad habitats: the field, the wood and the seashore, though the first two habitats tend to overlap a little. There is no expectation that the book should be read in order, and you are duly exonerated from the guilt that I always feel when treating a book as a bran tub.

The book is replete with words that may not be familiar to you, so (at the obvious risk of annoying those who know them) I include a brief definition alongside each one. No apology is needed for the Latin names that inevitably come with the territory. Units are a problem in any book that deals with historical agriculture as the metric system sits uncomfortably with the furlongs, bushels and pounds that were used. It has been necessary to simply muddle through. All the references to sources used in the writing of this book are on my website: https://www. foragerscalendar.net/spottersguide. Many of the photographs for this book were taken in 2021, so perhaps you will forgive the preponderance of images from the southwest of Britain where I live.

All of us question what we see when we drive along or take a walk. If we do not, then I think we should. I despair of those few for whom the countryside is no more than a pretty place to view from a distance or somewhere to take exercise. Should they be unfortunate enough to find themselves in my company, I have the inexcusable habit of trying to change them, to slow them down and encourage them to question what they then see. I do not intend to change you, dear reader, but I hope at least to inform and, in explaining some of the mysteries we encounter on our walks, to show you how very deep they can be.

John Wright
2021

The Field

Water Meadows

Amongst most pleasant Meadows, maine of which of late yeares have been by Industrie soe made of barren Bogges.

Thomas Gerard, *Survey of Dorsetshire, c.*1630

Apart from coastal areas and the occasional bog, the British landscape is remarkably dry. With the fairly high rainfall typical of a northern hemisphere land mass at the western edge of an ocean, it should be much boggier. Once, it was, with low-lying land, especially near rivers, being no-go areas for people and unusable for agriculture. But the efforts of farmers and landowners over thousands of years have banished most of our wetlands, with those that remain now national treasures of biodiversity.

River flood plains were long ago cleared of wet-tolerant forest and partially drained, claiming them for pasture or meadow. This land was usable, but it was still generally wet and of poor productivity except in high season. Water Meadows, usually seen from valley-bottom roads, and familiar as conspicuous, geometrically arranged ridges and channels in flood plains alongside their supplying rivers, seem to go against the flow of agricultural improvement by making the land wetter. However, this was not a perverse enterprise but one designed to raise soil temperature to give an earlier and more productive growth of pasture. Additionally, in Water Meadows the water levels are controlled, and are only high for a short period of time; most of the year a Water Meadow would be drier than the pasture that preceded it. A flood plain it still was, but one that was controlled.

In his *Treatise on Watering Meadows* (1792) George Boswell decries the waste of so much land that could easily be improved (in an agricultural context 'improving' always means 'improving yields'):

> *Every gentleman who has travelled thro' the different counties of the kingdom, with a view to its improvement, must have observed the very great quantities of unimproved, boggy, rushy, wet land, that lie almost every where near the banks of rivers and lesser streams, that seem to have baffled the skill of the possessors.*

Boswell, a tenant farmer in Dorset and a great agricultural innovator, then goes on for the rest of the book encouraging landowners to install a particular version of Water Meadow on their unproductive pasture and telling them how they might proceed.

The aim of flooding a flood-plain meadow is to provide early grazing, followed by a good crop of hay. But how will it help to flood a meadow that is already wet? The answer is that it raises the temperature of the soil. Experiments have shown that Water Meadows maintain a minimum of 5°C, the lowest temperature at which grass can grow and seeds germinate, with the additional and considerable advantage that the ground never freezes. Furthermore, the water brings with it sediment and nutrients which have washed from the soil upstream and, since many Water Meadows are in the chalk lands of the south, carbonate ions to raise soil pH and prevent acid conditions.

Water Meadows have been around since the late Middle Ages. The early system, if 'system' it can be called, was termed 'floating up'. This is an odd term as there is little about flotation associated with flooding, but I encountered an early spelling of 'floated' in this context which was 'flodded'. It seems that the words are divergent cognates, and I attempted to confirm my suspicion with online dictionary searches. While I am convinced that I am right about this, my researches were cut short when I discovered that 'flodding' had acquired another meaning altogether. I beg you not to look it up.

'Floating up' is not a sophisticated process – the farmer simply dammed the river downstream of the meadows he wished to flood, then dismantled the dam in March. There are few records of this method and almost no archaeological remains, as they would not leave any of note. So it may be that the meadow was only dammed intermittently to keep the grass fairly

Water Meadows near Dorchester, Dorset

fresh. Certainly, it was a risky practice; with the oxygen levels dropping, the grass would suffer.

Worthy of note is a type of Water Meadow known as a 'floating down' or 'catchworks'. Against all expectations, these were on the side of a hill. The water would arise from a spring, a small upland stream or a large pond, and run in a channel down the side of the field to be watered. More or less horizontal channels which run across the field would be fed from this feeder channel, and whatever had not spilled over into the field (as it was meant to do) was run into further Water Meadows on the flood plain. It was customary with this system to guide the water through any convenient stockyard prior to its journey down the hill so that muck could be incorporated. Human waste also made its way into the meadow, but these were times when nothing was wasted.

Both the watering and the aromatic fertilisation would have had a dramatic effect on the grass in the meadow. I have seen this effect on a couple of occasions, though neither was welcome. The farm on which I once lived consisted mostly of herb-rich chalk downland. Unfortunately, there was a farm on the hilltop above one of the steep chalk slopes which kept pigs in open fields. All very commendable, and I am sure that the pigs were very happy right up until just before the end. The problem was that the farmer was not very good at preventing pig slurry running down the hill onto the herby grassland below. What had once been a short, springy, rabbit-nibbled turf with dozens of species of plants and hundreds of species of sometimes rare invertebrates became a mass of lush grass, three feet high. This is called 'eutrophication', meaning 'good growth', and is the bane of biodiverse habitats everywhere.

Catchworks went out of business in the seventeenth century after a 400-year run and can now be seen only rarely and in relict form as incongruous, partly horizontal ridges on hillsides, chiefly on Exmoor. By the beginning of the seventeenth century a much more complex and practical system was coming into favour, the familiar 'bedworks'. This system has developed over the centuries, but the general idea, and simplest form, is as follows. A weir and sluice gate are built to control the river. Downstream, just below the weir, a channel called the 'main carrier' is dug (and banked where necessary) to form a loop that will re-join the river further down, at which point there would generally be another weir and sluice gate. At right angles to the main carrier a series of shallow banks are built with 'subsidiary carriers' on top. The sloping sides of the banks are known as 'panes'. Between the banks there will be a drain which empties, eventually, back into the river. The ridges (two panes plus a carrier) are anything from 3 to 15 metres wide, and about half a metre high. There are endless arrangements and variations on this, such as the banks being in a herringbone, and aqueducts used where there was a tricky situation. Sometimes it was necessary to add culverts or even little bridges so that the worker could cross areas of water without getting wet. Some of these constructions can still be seen today and, if the land has been ploughed, they may be the only obvious sign that a Water Meadow was ever there.

Once the sluice gate was opened, the water would flow down the main carrier, into the subsidiary carriers and flow over the side. The aim was to have an ever-flowing 'sheet' of water, no more than 25 millimetres thick, running over the panes. Once the Water Meadow was built, permanent pools would sometimes be problematic and were filled with the sediment that needed to be cleaned from the various drains. Holes in banks would be repaired with turfs.

Bedworks were first created in the southern, chalky counties of England, where the climate and preponderance of broad valleys were eminently suitable for the practice and a tradition of sheep-keeping well established. Dorset was the driving force in this, the Water Meadows constructed *c*.1600 at Affpuddle (the ones managed by George Boswell 200 years later) being regarded as the first bedworks in England. In the late eighteenth century, of the 775,000 acres of the *old* county of Dorset, 50,000 acres were occupied by Water Meadows.

There are a few in East Anglia, though they generally dispense with banks and have only drains, the famously flat landscape not providing a sufficient head of water for anything else. They are fairly rare in the north because the topography, poor drainage and low winter temperature of rivers do not encourage them.

To understand the workings of a Water Meadow completely, it is necessary to understand the agricultural practice it was meant to serve. Along the chalklands of the south this was 'sheep-corn husbandry'. In the particular case where Water Meadows were used, this is how it went. In December the Water Meadows would be flooded as already described. When first flooded, the ground was reputed to sing from the sound the water made as it filled a million wormholes! The sheep were kept on the downs, fed with last year's hay and whatever they could find in the fields. In March the sluice gates would be closed, and the meadow allowed to dry for a short period. The grass would continue to grow quickly and well in the relatively warm soil. The sheep (though it was usually just the ewes and lambs) were then brought to the meadow for a few hours each day, often feeding in pens which were moved every day. This was to ensure that the grass was conserved by not allowing the sheep to eat indiscriminately all over the meadow, and so that they could easily be gathered for the return journey. Once off the meadow, the sheep would be folded again, this time on land destined to be sown for corn. This could be on the shallower valley sides or high up on the tops of the downs. Here they would fertilise the land in the normal course of events, bringing nutrients back from the waters that had run off the hills and into the river.

From the beginning of May the sheep were released onto the permanent grasslands of the downs. In late June or early July the hay in the Water Meadow would be cut. Sometimes a second hay crop could be had, drought being a rare problem as the meadow could usually be flooded again at will: 'damping', as it was called. Then the cut meadow would be turned over to sheep again or to cattle. Cattle could only be pastured when the ground was sufficiently dry to prevent them turning it into a muddy mess. This may all seem like a way of creating sheep, but in truth it is the corn that is the primary concern. Barley, wheat, lambs, wool and some dairy for domestic use were the chief products of this method of farming, and in that order of importance.

There is something very appealing and efficient about this idyll, with sheep being moved twice daily, shearing time, haymaking and fireworks near the party tree. But no agricultural practice is immune to changing times. Cheap fertilisers and cheap food imports made the Water Meadow an economic fossil, and after its heyday in the mid-nineteenth century, when sheep reigned supreme and the meadows flowed with river water, Water Meadows slowly declined. Apart from a handful of living museums, the practice has died out completely. Many of those that were left as merely rough pasture have disappeared under the plough, 'improved' drainage making possible a previously inconceivable idea.

If Water Meadows ever do come back, then I will be applying for the job of tending them and operating the sluice gates. It comes with a great title: I will be a 'drowner'.

Sheepfolds and Sheep Stells

And Abel was a keeper of sheep.
   Genesis 4:2

The countryside abounds with odd-looking structures with no immediately obvious use: partial enclosures miles from anywhere and lengths and patterns of wall that go nowhere and enclose nothing. Most of these are designed to control and protect livestock, and it is mostly sheep that need to be protected and controlled.

The folding of sheep has always been integral to farming, though there are effectively two types. One is the straightforward matter of keeping them safe and all in one place at night; the other is to confine them temporarily to any one area for feeding, shearing, awaiting transportation and so on. The latter are transient in nature and leave no memorial except in the written record. In the past an area in a winter field of brassicas or a summer field of sainfoin or vetch would be fenced off with hurdles. There could be twenty-five hurdles on a side, which would allow 200 ewes on a good crop for one day. They would be moved to another fold the next day. The sheep were all fed and the soil was well fertilised. Such picturesque sights are seldom to be seen, but the practice lives on with both sheep and cattle controlled by the infinitely less charming electric fence.

The more familiar Sheepfold is a permanent structure of stone. Thousands of these can be found across Britain, though naturally they are almost all located where stone is readily available. There are no permanent sheep pens in the chalklands where I live, though they are found on the harder limestones of Purbeck, 25 miles to the east. As with pounds (see p. 33), circular Sheepfolds are the most common, though a great variety of forms are known. It is, just about, worth mentioning that Sheepfolds do not have a roof; a Sheepfold with a roof would be a stall, barn or shed.

A very impressive survey, carried out by volunteers confined to quarters during the difficult days of 2020 and published in August of that year, mapped and described hundreds of agricultural structures on the Lammermuir Hills to the southeast of Edinburgh. The work was undertaken by volunteer archaeologists, and almost entirely online using various sources such as Google Earth and LiDAR. They discovered 860 sites of interest, among them over 300 Sheepfolds. There were also a large number of other enclosures, some of which may have been roofed over, which were also likely to have been used to keep sheep safe and protected from the elements.

A handful of the folds that they recorded were rectangular, but the majority were circular. A few were more complex: there were 'keyhole' folds, which are circular with a rectangular structure at one end, and two 'spectacle' folds – circular folds linked by a short

wall, the entrance to both circles being at the join of fold and wall and on the same side. The typical diameter of a circular fold was around 18 metres, with a range of 8 to 25 metres. Although most of the folds were made from stone, around 40 per cent were made from turf. These latter had not been extensively mapped before as they are not easily spotted among the vegetation, but LiDAR technology made them clearly visible.

Conspicuously absent from the open, rolling and heathery vistas of the Lammermuir Hills are dry-stone walls. Walls act as a physical division, but they also provide shelter. The folds and other enclosed structures obviously provide shelter, but there is the 'halfway house' of Sheep Stells, twenty-eight of which were recorded. These are sometimes straight walls that appear 'stranded' in that they are attached to nothing. Other stells take a variety of shapes and are named accordingly: 'Y', 'X', 'V', 'T' and 'C'. Stells are the permanent equivalent of the windbreaks every self-respecting Britisher takes to the seaside, excepting that, being fixed, they cannot be positioned exactly against the current wind direction. The various shapes of stells go a long way to countering this drawback because there is always *somewhere* inside or outside that protects from the wind.

With features such as those catalogued above, it is tempting to think of them as ancient; however, it was noted that wherever the ridge and furrow traces of runrig cultivation (see p. 72) were seen, any fold or related structure clearly overlaid the field system. This makes it certain that the folds (and sheep) came later, probably in the late eighteenth or early nineteenth century.

Elsewhere in Britain, where stone walls were part of the scenery, Sheepfolds would often be built against a convenient wall. This often took the form of a semicircular wall resulting in a 'D'-shaped fold. Some folds would run at right angles out from a wall, then curl around to form a spiral, the entrance being the gap between the wall and part of this spiral. The wall would guide the sheep into the fold, and indeed some folds in open countryside have a short length of wall built tangentially to a circular fold for the same purpose. From above it looks like a minim on a musical stave.

The most beautiful of the Sheepfolds are those that have simply accumulated. These include multiple enclosures which show every sign of being added over time as the need arose. They can even include a small stone hut, complete with an open fire, which would have been a great boon during cold weather for the shepherd, and a life-saver during lambing. These are romantic structures, often alongside a lichen- and moss-covered wall, nestling in a landscape of which they seem such a natural part. It takes a strong materialist heart not to dream of moving in straight away.

Sheepwash

With whistle, barking dogs, and chiding scold,
He drives the bleating sheep from fallow fold
To wash-pools, where the willow shadows lean,
Dashing them in, their stained coats to clean;
Then, on the sunny sward, when dry again,
He brings them homeward to the clipping pen,
Of hurdles form'd, where elm or sycamore
Shut out the sun – or to some threshing-floor.
    John Clare, 'June'

Apart from the more colourful rare breeds and those that have been brightly graffitied with identification and tupping marks, sheep appear unquestionably white when seen from a distance.

A close encounter, however, reveals them to be extraordinarily grubby creatures suffering from a permanent bad-hair day. In the past, farmers wishing to send their wool to market would nearly always wash their sheep a day or two prior to shearing. This resulted in a small (3 per cent) increase in the value of a fleece, but it would have been lighter in weight after all that washing, and I suspect it was much more about pride in one's work than financial advantage.

Washing a flock of sheep is no small task, even with dogs to keep things in order. The traditional and obvious way was to use a river directly, though a suitably convenient and safe spot was always chosen and may have been used every year for centuries. Some places for which no remains or records of a constructed Sheepwash are known could still be called 'Sheepwash' or 'Washing Pool'.

One simple way to achieve clean sheep is to erect a length of fencing down a stretch of river and drive the sheep along the watery path thus created. A cheerful report of a slightly different method practised at Bromyard in Herefordshire appeared in a newspaper in the early twentieth century, penned by 'The Amateur Shepherd'. The sheep, along with several helpers and a cask of cider, were led to the river's edge. Here there was a small drop into a 5-foot-deep, semi-natural pool. The sheep would be launched, one at a time and upside down, into the pool, where they would helpfully splash about for a while before embarking on a breaststroke – well, doggy-paddle anyway. Two men, wielding sticks to each of which was attached a partial hoop which fitted nicely onto a sheep's back, stood on rocks either side of the pool. It was their job to control the sheep and ensure they received a good dunking. The sheep would then be led by other men across the (presumably shallow) river to the far bank.

A more robust approach can be found in Bridgend, south Wales, at the elegant four-arched stone bridge across the River Ogmore. Here two sheep-sized 'doorways' can be seen in the parapet, through which sheep were encouraged to exit (they were pushed) and fall about 5 metres into the river. Helpers were at hand to lead the traumatised but *clean* sheep back to dry land.

Various more complex arrangements have been made to ensure a clean fleece. Usually they are found near a river or stream but can also be up in the hills and fed by a spring. Some were just stone-walled watery runs of about 10 metres, as at Scarcliffe in Derbyshire. The commonest, however, is a round (or, less frequently, rectangular) stone or brick pit, with water running in through a culvert or sluice gate and then drained through a wide channel which would also allow the passage of clean sheep. Circular specimens are about 10 feet in diameter and 5 feet deep.

Such Sheepwashes are fairly common and rather splendid-looking things, though, sadly, many have been lost to neglect over the years. A survey of the Sheepwashes in the Cotswolds (most of them in Gloucestershire) found evidence of sixty-nine, only twenty-eight of which were still intact. The rest were classed as 'derelict', 'demolished' or 'ruined'. Years ago I noticed a larger, semicircular stone wall alongside the River Frome twenty minutes from my house. It was effectively a river wall, the top level with the surrounding water meadow, and had a sluice gate at the top. I now suspect it to be an unrecorded Sheepwash, though whether or not there was another half I cannot tell (perhaps the other, entirely natural, bank was considered sufficient). If the other half had existed, then it was an exceptionally large Sheepwash, at 20 metres in diameter.

The washing of sheep, or at least washing them using the above-mentioned techniques, is now largely a thing of the past. Most Sheepwashes had fallen into disuse by about 1920, though at least one was used as recently as 1960. However, it is thought that the latter was not for washing but for dipping, which is something else entirely.

There are remains of old sheep dips around the countryside (some of them, as we have seen, being commandeered Sheepwashes); they are generally easy to recognise by their long and most often rectilinear shape (in one end, out the other) and their location away from running water. They are much later than most Sheepwashes, starting their sometimes ghastly career in the middle of the nineteenth century. Sheep, weakened as they are from selective breeding and living in places for which they were not 'designed', suffer from a variety of unpleasant diseases and drop dead at a moment's notice. To eradicate at least some of the diseases they were, and still are, briefly bathed in something unpleasant. Unfortunately, arsenic compounds were among the first products devised (by William Cooper in Berkhamsted in the 1850s), but anything sufficiently poisonous to kill mites, ticks (scab), blowflies and lice is likely to kill a lot of other things as well. With arsenic, it could be people. Safer and more sophisticated chemicals have been used since, but it was barely fifteen years ago that organophosphates were banned or strictly regulated.

Pounds

All we like sheep have gone astray.
    Isaiah 53:6

Just around the corner in my village is Pound Piece. Here there are a few modern houses, the surgery and a very large rock, set back a little from the pavement and looking like a meteorite that hadn't been in much of a hurry. It is known as the Millennium Stone, marking the turn of the last century and also the site of the village Pound, now long demolished. My village is not alone in having, or once having had, a Pound; very nearly every village in the country boasted one, and many are commemorated by a road name, a house name or, of course, a rock.

A Pound is simply the place where straying animals were taken, to be reclaimed by their owners later, and, less frequently, a place to keep animals that had been confiscated in lieu of rent and were awaiting payment or, failing that, sale. It seems strange that so unusual a happenstance as roaming animals required so universal a solution, as it is relatively rare to find stock wandering aimlessly around the countryside. It was not always so. Sheep were moved very frequently – from hillside to low pasture and water every day – and cattle may have been milked in the village.

If stock was taken to market, it had to walk there. Also, compared with a very few decades ago, sheep and cattle are unusual sights now, with sheep farming a rarity in the lowlands, at least, and cattle and pigs often bred indoors.

Lost animals were once a big problem, one that was shared by both those who lost them and those on whose land they were found: the latter were feeding someone else's stock. Pasture was a high-value commodity which was jealously guarded, so stock was sometimes 'lost' on purpose. The unscrupulous, or just the desperate, would allow their animals to stray onto land owned or leased by another, and if that involved moving a fence, then so be it. With pasture once being 'in common', straying animals were a constant problem and physical enclosures were rare. The 'waste' – land that was used for the collection of firewood, picking wild berries and rough grazing – was in common ownership, but each household in the manor would enjoy different rights of pasture to various parcels of land at various times, which naturally created opportunities for stock to be in the wrong place.

Anyone could take stray animals to the village Pound, but a responsibility was then placed on him or her according to statute. A Mr Richards, who owned a farm on Mynyddislwyn Mountain, in north-west Gwent, in 1888, found sixty-six of his neighbour's sheep helping themselves to a field of oats and the June grass from his meadow. He took them to the village Pound and put them in the hands of the pinder (the person who managed the Pound). When the errant farmer came to claim his sheep, Richards charged him £6 12s. This was duly paid, but the owner of the sheep was dissatisfied with the way they had been kept and charged Richards £6 12s right back. The ensuing dispute ended in court proceedings, with allegations of vindictiveness bandied about and the professionalism of the unfortunate pinder impugned. The case was settled against Richards.

The owner would retrieve his 'lost' stock on the payment of a fine. This was to cover the cost of feed, maintenance of the Pound and the wages of the pinder (Pounds were frequently called 'pinfolds'). Not everyone was inclined to pay the fine, and some would rescue their animals at the dead of night: 'poundbreach', as it was called in parts of the country. Pounds would be locked, and the walls were generally at least 5 or 6 feet high, as anything lower would encourage such sneaky activity. In later times, advertisements were placed in newspapers for anyone who seemed to have forgotten they were missing a few animals:

> City of Hereford. Palace Pound. Notice is hereby Given, There is now in the Palace Pound FOUR DRY SHEEP – The Owner may have them again by applying as above, and paying the Expenses.

If no one claimed the stock, they were sold directly or at auction. A dramatic instance of an auction occurred in Ireland in 1826, when that country was unhappily under the rule of the British.

It was, however, for the sale not of unclaimed stock but of confiscated stock. A newspaper article sets the scene:

> *On Monday last, pursuant to notice, the Rev. Arthur Preston, rector of Kilmeague, commenced selling by auction various lots of cattle and sheep, distrained for arrears of tithes. Around the Church, and the village pound, was posted a strong party of police, in number about 500.*

Five hundred constables not being considered sufficient, the authorities also brought in the big guns, literally, with two squadrons of the 5th Dragoons, two of the 10th Hussars and several companies of the 5th and 42nd Infantry. Just to be sure, they situated a couple of brass 24-pounders and teams on top of a nearby hill. This may seem excessive, but the report tells us that 20,000 locals were in attendance, flooding in from nearby parishes, not to bid but to give support to those whose stock had been confiscated.

A mounted farmer roared, 'Never! No tithes!', and the chant was taken up by the crowd. The reverend and auctioneer were seemingly of the same mind and the auction perforce became a fiasco. A 'fat cow' valued at £4, went for 3s, barely 4 per cent of its real value, and the rest of the auction went the same way. Despite the clear potential for an all-out battle, the day passed by peacefully, the victory clearly that of the locals. A charming addendum to this story is that the rank-and-file British soldiers, marching back to barracks, and no doubt the sons of farmers or farm workers to a man, began chanting, 'No tithes! No tithes!' Social class often trumps nationality when it comes to oppression.

Village Pounds come in various shapes and sizes, circular being particularly favoured as no animal will then find itself crushed in a corner. Circular stonework is easier to build, and stronger, as a rectangular construction is naturally weak at the corners unless dressed and mortared stone is used. None, to my knowledge, ever came with a roof. Some, such as manorial Pounds and Pounds of well-to-do towns, can be grand constructions with high-quality dressed stone. Many of all types still survive, and most of these are within a village or just on the edge. With any village that has managed to grow into a town, the Pound can be effectively in the middle. There are some 'mountain pounds' in areas such as Cumbria. The purpose of these far-flung Pounds (or pinfolds) is not entirely clear, as without documentary evidence one way or another they are indistinguishable from Sheepfolds (see p. 26). Thus, on Ordnance Survey maps, a likely candidate will be called a 'pinfold' in one edition and a 'sheepfold' in the next.

As with most of the pastoral furniture in this book, the village or manor Pound is no longer in use, with the last reference I can find being for Hope pinfold in Derbyshire, where 300 sheep were impounded over the course of 1967. The charge was 2s 6d a head in summer and just 6d in winter, the difference representing the relatively enormous damage sheep can do when the fields are full of good grass and corn.

Marl Pits

Over the burning marl
    John Milton, *Paradise Lost*

To the south and east of Chester there is a gentle landscape of field and hedgerow. Pleasing as it may be, it appears very ordinary, almost tediously so. Nothing, then, to see here. But there is. A very large proportion of these fields have a small pond in them, very often right in the middle. Some of these ponds have dried up, most are skirted by tree and shrub, and nearly all of them are vaguely rectangular. With Cheshire not being known for its repeated droughts, or any indication that the population was inordinately fond of ducks, there seems no reason for them and certainly not for so very many. They are abandoned Marl Pits that have filled with water. Doubtless there were many more that have been filled with something more substantial over the years by farmers who had tired of an irritating pond getting in their way.

Since few people know quite what a Marl Pit is or is for, it is surprising to discover that digging a Marl Pit on your own land is one of the few things specified as being allowed in the thirteenth-century Charter of the Forest, and therefore one of our most fundamental rights:

> *Every freeman from henceforth, without danger shall make in his own wood, or on his land, or on his water, which he has within our forest, mills, springs, pools, marlpits, dykes, or earable ground, without enclosing that earable ground, so that it be not to the annoyance of any of his neighbours.*

Marl is a natural mixture of clay and calcium carbonate, the latter simply being limestone. Its texture is sticky and silty, with gritty bits. It has long been used as an improver of soils; Pliny the Elder, no less, mentions its use in Britain in the late Iron Age, and it continued – indeed increased – in use until it was replaced by burned lime and then milled lime when the canal and railway systems were sufficiently developed to carry bulk materials. The last pits were dug in the late nineteenth century.

The chief purpose of marl was to reduce acidity, but it would also improve the consistency of clay soils. While marl was often transported some distance, it seems extraordinary that the very thing a farmer needs is quite possibly just below the field on which it is required. The explanation, however, is straightforward. Soils are gradually depleted of minerals over time through leaching caused by rain, and removal with the gathered crops that absorb them. Yields decline.

Calcium is the key nutrient in rebalancing the soil pH and to find it conveniently available as marl in the bedrock below the afflicted field must have seemed like a godsend to worried farmers. Soils are made, in part at least, from the subsoil. The soil may accumulate other matter – wind-blown particles, organic matter – but it will also lose something through the leaching out of soluble inorganic compounds by rainfall. The chief effect of this is to lower the pH, resulting in a nutrient-poor, acid soil. The purpose of digging out and spreading marl is to bring the subsoil to the top, where it can raise the pH again. To be pedantic, marl is really the bedrock and not strictly the subsoil – subsoil is a pretty loose term but stops at the level where organic matter is negligible or zero. In addition to improving fertility,

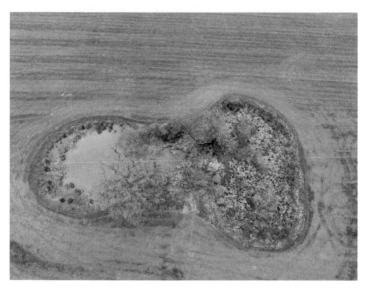

the calcareous component of marl binds the clay particles in the soil (flocculation) to make the texture more friable and allow drainage, aeration and better root growth.

Gervase Markham, writing in 1625, tells us how it was used:

After your first Marling [...] you plough not the ground either with deep or broad Furrowes, but fleet and narrow lest you cast your Marl into the dead Mould [...] Marl sendeth his virtue downward, and must be kept aloft and must not be buried in any wise.

In other words, marl was used as a top dressing.

Judgement would still be needed, as marls vary in their proportion of the valuable limestone. This can be anything from 35 per cent to 65 per cent, with the low and high values characterised respectively as 'poor-man's marl' and 'rich-man's marl'.

Marl is a common soft rock found throughout much of Britain, though the Mercia Mudstone is the largest formation. The latter is found as a large patch south-east of Liverpool (as we might expect), and most of the rest forms an untidy serpentine line from Middlesbrough to Sidmouth in Devon, with a large bulge as it passes through the Midlands. The bed also stretches north into southern Lancashire, which, like Chester, is noted for its Marl Pits. Many other geologically amenable parts of Britain sport Marl Pits, such as on the Wealden clays of Sussex and Kent. Ashdown Forest is fairly well known for them.

Records from the Chester area reveal that marl was spread during the autumn or winter and required a considerable hired labour force of five or six men (known, unimaginatively, as 'marlers') to dig it out of the ground. According to a letter to the editor in the *Athenaeum* in 1808, the work was carried out with all the ceremony that is due to a practice that had persisted for half a millennium: 'When a set of these men are hired to dig for marl in any particular spot, they choose from among themselves a leader of their body, who is entitled, by way of distinction, "The Lord of the Pit".' This character, straight out of fantasy fiction, would then harangue passers-by into giving him money, which, by tradition, they normally would. Once in receipt, the workers would down tools and dance in a circle, whooping and singing, 'Mr [enter name here] has given to my Lord and all his men a part of a hundred thousand pound!' The writer suggests that both money and workers were destined for the alehouse, though I harbour a suspicion that the writer penned his letter in just such an establishment. On the other hand, we still get Morris Men round our way and no one would believe they are real without personal and sober observation.

The pits are basically rectangular, with one end swollen and rounded – think of a keyhole. They come in many sizes, with the single pits still observable ranging either side of 15 × 22 metres. Several pits would be dug in a field, almost always adjoining a previous pit. Some of these neighbouring ponds are still distinct, while others have become single, large ponds. While they could be very deep, at 6 or more metres, an empty pit could be walked into because a slope descends from the square end until it reaches the round end, allowing a cart to back into the pit. Effectively, it was open-cast mining with straight access slope and rounded pit, but on a tiny scale.

The area south and east of Chester was once renowned for its high productivity, and the hard work of countless Lords of the Pit should be commended. These relict ponds (though they are always referred to locally as 'pits'), their agricultural capital now spent, have a new role as wildlife habitats, though a decreasing one as pits are lost to agriculture, urban development and, not least, the M53.

While the Marl Pits are long disused, many (a few thousand in the Chester region alone) are still to be encountered on a misty day, presenting a health and safety problem for the unwary:

Here lieth the body of Thos. Greenwood who died May 24 AD 1776
In ye 52 year of his age

Honest, Industrious seeming still content
or did repine at what he underwent
His transient life was with hard labour filled
And working in a marlepit was killd.

Mr Greenwood, who is buried in Ribchester churchyard, Lancashire, is not alone in having met his end in a Marl Pit. A quick scour through newspaper reports discovered many occasions when the dig collapsed on a worker, and other reports where people had drowned. In 1904 an eleven-year-old girl from Burslem, Staffordshire, was heard screaming for help, but the difficult terrain and deep water made it impossible to save her. I have never fallen into a Marl Pit, but a geologist friend of mine tells me that it is an occupational hazard and he seems quite proud of how many he has needed to clamber out of.

Tussocks

Green grow the rushes, O.
   Anon.

Some puzzles have haunted me for years, though frankly they would not have done so if I had taken the time to do a little research. One such was all about Tussocks. Tussocks are simple enough to understand – clumps of certain grasses, or grass-like species, that stand out from the surrounding vegetation – but there can be much more to them than that.

The mystery (to me, anyway) was a grassy area of the New Forest: Balmer Lawn and neighbouring grasslands near Brockenhurst, a place I visit twice a year. It is a massive 1,000 acres in extent and made up of hundreds of thousands of evenly spaced, nearly identical, flattish and very slightly oval grassy bumps. Each bump appears to have been shaved by a master barber, the whole looking like a gigantic art installation.

Anthills might have been a possibility, but few ants could survive the frequent flooding and generally wet nature of the soil. Unusually well-preserved and abandoned molehills that have grassed over also come to mind, but no, not molehills. Tussocks, then? Well, Tussocks don't usually look like this, and the bumps are largely devoid of Tussock-forming grasses.

This is a very special situation in that it is a very special habitat, and one that you are unlikely to come across. But such special situations occur all over the country (though they might not be so rare as this one), and you are bound to be confronted with many on your own perambulations. Thus the story of Balmer Lawn is exemplary in the solving of such puzzles, and in the understanding of how plants interact with each other, with their habitat and with man.

The entire Balmer Lawn area is exceptionally flat and almost level, dropping only 3 metres in its kilometre length. A stream runs through it, and there are many connecting drainage ditches. Broadly speaking, it is classified as a 'streamside lawn', of which there are very many in the forest.

Ignoring the roads, buildings and people, the New Forest feels quite primeval, but without the hand of man it would be almost entirely a dense and very wet forest. There would probably be open areas of heather and grass, maybe even a streamside lawn, but the one that constitutes Balmer Lawn is man-made. Drainage has always been an essential component of much agricultural activity and a corresponding problem for conservationists, but sometimes a drainage scheme will result in a wonder.

A not very detailed map from 1814 shows no drainage ditches and no stream, though one from 1869 does show part of the stream. Apart from the maps, the fact that it has long borne the name 'Balmer Lawn' indicates that it has long been grassland,

The unusual tussock-mounds of Balmer Lawn

though there is no doubt that it was very wet grassland. In such a situation, various tussocky species of grasses and sedges will take hold. One of the grasses is Purple Moor-Grass, *Molinia caerulea*.

This is a beautiful, statuesque grass with a mixed reputation in ecological circles. As the expert on peat bogs and their restoration Dr Roger Meade has said: 'This noble and majestic grass stands accused of crimes against ecology.' The sin of Purple Moor-Grass is its habit of turning restored peat bogs into monocultures of itself, and its general tendency to take over wherever it finds a habitat to its liking.

Purple Moor-Grass can be seen as grand, tussocky vistas throughout the damper uplands of Britain but also in the wetter lowlands, with a substantial swathe through Dorset east of Dorchester and Hampshire west of Southampton. A large area of the latter population is in the New Forest. At heart, the strange mounds on Balmer Lawn are Tussocks of Purple Moor-Grass or, more accurately, Tussocks *originating* as Purple Moor-Grass.

In the middle of the nineteenth century a drainage scheme was devised for the area. It, and other schemes, were paid for by the Southampton and Dorchester Railway Company at enormous expense in exchange for permission to build their railway line across the forest. It was supported enthusiastically by the investigating officer, Mr Josiah Parkes, who, coincidentally, planned to establish a drainage-tile factory at Brockenhurst. The 'Tilery' was later converted into a house, which is still standing. The tiles were made, laid in the ground and the ditches duly filled. The open ditches that are visible now are additional and much later drains.

Although Balmer Lawn still floods in the winter (and I have frequently found it to be inches deep in October), for much of the year it is dry enough to allow a picnic without an industrial picnic blanket and quite often it is bone dry. The dramatic change in water levels caused by the drainage scheme was reflected in an immediate change to the ecology, with the tall Tussocks of *Molinia* and sedge that were once there suddenly a thing of the past. Not only is the land drier, but it is also more accessible to grazing animals, which was, after all, the whole point of draining it. In the case of Balmer Lawn, it is New Forest ponies that are the barbers and the ultimate cause of this striking landscape.

Comparing grazed riverside lawns with those that are ungrazed, my saviour in this puzzle, P. J. Edwards, found that while *Molinia* was dominant in the latter, it was absent or much reduced in the former. The *Molinia* Tussocks had become a scaffolding within which other plants could grow. These, Edwards notes, included less flamboyant grasses and sedges such as: Nardus, *Nardus stricta*; Sheep Fescue, *Festuca ovina*; Heath Grass, *Danthonia decumbens*; and also Carnation Sedge, *Carex panicea*. Also notable was the large amount of sphagnum moss on the mounds and, within relatively recent grazed mounds, dead remnants of *Molinia*. My mystery Tussocks were once of *Molinia* but have been transformed into a home for other species. One other mystery was solved by Edwards. The mounds are now largely made up of silt, held in place by the vegetation. The silt was deposited in the former Tussocks during flooding, and the mounds acquire an oval shape from the further deposition of silt in the slack water immediately downstream of each mound.

The short list of acquired grass species above is a general result from the surveys of small areas of riverside lawn and only an indication of what has happened at Balmer Lawn. However, in his book *Flowers of the Forest*, Clive Chatters writes about Balmer Lawn and relates how species-rich this unusual community has become. There is a charming subset of botanists devoted to the dandelions. Most people think that there is only one dandelion, but how wrong they are. There are 230 'microspecies', most of them asexual and thus clonal. Balmer Lawn is famous for them, and Chatters writes excitedly about a hybrid (hence 'most') between *Taraxacum palustre sensu stricto* and *T. ciliare*. He expresses a hope that it will become a recognised microspecies reflecting its only known location: *Taraxacum balmerense*, perhaps. In addition to the admittedly minor sport of dandelion-spotting, there are many other flowering plants found on Balmer Lawn, over a hundred, putting it in the top ten sites for plants in Hampshire. Finally, it appears that ants do occur in some of the mounds – the drier ones, no doubt. They are the common anthill species the Yellow Meadow Ant, *Lasius flavus*. No moles, though.

There are hundreds of individual classes of plant community in Britain, most of which have been described and catalogued, albeit using impenetrable names such as '*Ulex gallii – Agrostis curtisii* heath, *Scirpus cespitosus* sub-community'. Wherever you find a plant growing in a wild or semi-wild habitat, it will be part of a classifiable community. We may describe a walk as being across a few fields and through the woods, but this may encompass any number of separate communities: rush pasture, lowland dry acid, lowland meadow and pasture, lowland mixed oak and ash, lowland dry oak and birchwood. In fact, most communities have subdivisions, so the five mentioned here could easily be fifteen. The combe valley near the dew pond mentioned in the next chapter contains no fewer than ten!

There is great satisfaction in knowing precisely what is going on around us. Many with a good grounding in the identification of plants will enjoy counting them off as they go, finding with delight a few they have not seen for a while or never seen before, and admiring the scene. But to gain a deeper understanding it is necessary to appreciate both how these plants work together with invertebrates and fungi, and how they reflect the local climate, topography and soil.

Hundreds of habitat types are given snappy names like the one already mentioned, all described and explained in minute detail in a series of books which include such titles as *British Plant Communities, Volume 2: Mires and Heaths*. Other volumes cover grasslands and montane plant communities, woodland communities and those of the seashore. They make fascinating reading, though it takes me an hour to read any one page. These books are neutron-star dense. First published in 1992, they seek to describe the many different plant communities of Britain and classify them.

It is perfectly possible to work these out for yourself, provided you are very good at plant ID, are prepared to spend an hour negotiating the question-and-answer session that is the 'key' and have absolutely nothing better to do. I am not particularly good at it, but you might be.

Does Balmer Lawn fit into any recognised community type? Just about. It is classed as 'M24c *Molinia caerulea* – *Cirsium dissectum* fen meadow, *Juncus acutiflorus* – *Erica tetralix* sub-community', where 'M' stands for 'mire' and 'c' specifies the sub-class. I mentioned this to a friend who spends much of his time surveying ecosystems and who knows Balmer Lawn very well. He told me that it wasn't as simple as that.

Dew Ponds

Jack and Jill went up the hill to fetch a pail of water.
Traditional

My first home in West Dorset, some forty years ago, was a seriously isolated farm cottage high up on the chalk. Situated as it was just below the summit of a west-facing slope, weather was an issue from day one. It was late April when we moved in. Snow fell heavily until it lay 5 inches deep. Over the next few days it was melted by a heavy rain which fell horizontally and, after some determined effort, began to run down the *inside* of the western wall of the living room.

Chalk downland, the one-time favoured haunt of Neolithic man, has always been a difficult place to live – not so much from the trials of snow and driving rain as because of the water disappearing as soon as it melts or falls. Typical of such landscapes, the valley bottom 200 feet below the cottage bears no stream or even a ditch, and water seldom runs from the hills but instead sinks where it falls through the thin soil and into the porous and fissured chalk. There are no springs, no rivulets of water appearing after rain, no natural ponds. It is a very odd sight to those more familiar with a less porous landscape; surely, they think, there must be a stream *somewhere* in this valley.

There are, however, *un*natural ponds – Dew Ponds. On the farm there were, and still are, two: one near the farmhouse, one at the top of the east-facing slope on the opposite side of the dry valley. I had never encountered Dew Ponds before and, with no stream or spring to fill them and sitting as they were on porous chalk, I could not understand how they could exist. My friend Reuben, a local who knows about these things, explained that they gather water by condensing dew from the air. Thus replenished, they were able to remain full, even when rain failed to fall, retaining water in severe drought when ponds in the river valley fail. If you ever wondered why Jack and Jill went *up* the hill, this is why.

There is much speculation, and little evidence, on how long such ponds have been built on the chalk uplands, with Neolithic vintage being championed in a beautifully bound book of 1910, *Neolithic Dew-Ponds and Cattle-Ways*, by Arthur John

Hubbard and George Hubbard. Its argument is based partly and unconvincingly on the observation that Dew Ponds, and depressions in the ground that may once have been Dew Ponds, often occur near Neolithic features. A much stronger argument, but still not proof, is that if people once lived and farmed on chalk hills, they must have obtained their water from somewhere and hit upon the idea of Dew Ponds (or at least ponds) by noticing how puddles form and persist wherever the otherwise absorbent ground is heavily trampled ('puddled', as it is called) by stock.

Although the modern, if still uncertain, view that they originated in the Bronze Age seems reasonable, how continuous their use was depends on how the land was farmed: Dew Ponds are required for pastoral farming but not for arable. Agricultural practices changed back and forth over the years under the influence of economic and political events, notably the progressive enclosure of the land, and Dew Ponds came and went.

Generally, it is difficult, and usually impossible, to date a Dew Pond, but sometimes there is a strong clue. A good example is Dorset's highest village (at nearly 700 feet), Ashmore, which is conspicuously centred around a large Dew Pond. Most villages date from the Saxon period, and for Ashmore this is certainly so, as it is recorded in Domesday as Aisemere. This name is from the Old English *aesc* ('ash') and *mere* ('pond'), thus meaning 'pool where the ash trees grow'. Ashmore is miles from the nearest river, and the entire area, chalky Cranborne Chase, is scattered with high Dew Ponds. Another pond, known as Oxenmere, overlooking the Vale of Pewsey, was recorded in a Saxon charter of AD 825, so there is no doubt that Dew Ponds are truly ancient features of the landscape.

Several hundred still exist in various states of repair in Britain. Nearly all are on the chalk of the south, Yorkshire, the Lincolnshire Wolds and the White Peak in Derbyshire. Roaming

Next to the Hell Stone, an algae-infested Dorset Dew Pond

gangs of men would build them for farmers, and the last dedicated Dew Pond maker I can find was a Mr J. Smith, 'Dew pond and lake maker', from Market Lavington in Wiltshire. On the death of his father in 1937, he assured his customers in a newspaper advertisement that 'I am still carrying on the work which has been in our family for 251 years'.

But still they are made and restored, with the various projects concerned more with nature conservation than an upland water supply, something that is now provided from boreholes via pipes and troughs. A Dew Pond will greatly increase the biodiversity of any grassland area. The one near the farm cottage once had a hundred mating frogs in it, and I always approached the other pond with caution because of the inevitable and sudden uprush of disturbed ducks. Ducks, frogs and damselflies are always welcome, but better still is the Great Crested Newt, which has found a refuge in some of these ponds, notably in Kent.

With their purpose and antiquity more or less established, an explanation is needed of how, and even if, Dew Ponds stay reasonably full when seemingly better-situated ponds do not. The modus operandi offered by my friend Reuben was tentatively introduced by a Dr Christopher Packe in 1743. He mentions a large pond on Collier's Hill in Kent, noting that it had no obvious means of renewal beyond direct rain. He seems to be at a loss for an explanation but does at least mention 'condensation of elevated vapours'. Thomas Hale, in his excellent *A Compleat Body of Husbandry* of 1758, mentions what are probably Dew Ponds with the observation that 'Ponds on the tops of hills are not uncommon', but makes no rash claims about their hydrology.

The best-known early observation of Dew Ponds is that of Gilbert White in 1776. Writing about those found above his home in Selborne, he mentions one which was known as Wood Pond:

> *Now we have many such little round ponds in this district; and one in particular on our sheep down, three hundred feet above my house; which though never more than three feet deep in the middle, and not more than thirty feet in diameter and [...] containing not more than three hundred hogsheads of water, yet never is known to fail, though it affords drink for three hundred sheep, and for at least twenty large cattle beside.*

He also suggests that dew may be the saviour of these ponds.

Thus an explanation for Dew Ponds was born, or at least first intimated, by Dr Packe. There followed a widespread consensus that these ponds seldom dried out, and that they worked courtesy of dew condensing onto the cold waters of the night-time pond. But is this idea – one that *seems* to come complete with a self-explanatory name – correct?

In fact, the name 'Dew Pond' is not recorded until the early nineteenth century, and they were probably just called 'ponds' or 'sheep ponds' by those who required them for their livelihood.

Designations such as 'mist pond', 'cloud-pond' and 'fog pond' have also been employed, but no early record of these exists either.

Attempts to prove or disprove the theory began in 1837 with Henry Pool Slade (a man nominatively destined for this project), who dismissed it, and the controversy continued with eccentric enthusiasm well into the twentieth century. There were contemporary reports of shepherds claiming 1 to 2 inches' increase in water level overnight, even though no rain had fallen, so it should have been fairly straightforward to at least ascertain whether or not dew had a hand in matters or even that there were any matters for dew to have a hand in.

Not so. A series of books, articles and papers appeared which cast nothing more than shadow and confusion on the issue. Most of these resemble whodunnit stories with tediously impenetrable plots and a perverse determination not to tell the reader who, indeed, dunnit. Nevertheless, some facts and opinions shine through the murk and are worth relating.

How dew behaved was gradually elucidated in the nineteenth century, and it transpires (sic) that it is not deposited on anything that has a higher temperature than the water-vapour-laden air from which it may condense. Rather, vapour generally evaporates from the warmer surface and is sometimes seen condensing as a mist and rolling gently down the hill – a Dew Pond in reverse. Liquid water, such as that in a pond, will hold a great deal of heat, which it releases but slowly and which is always kept relatively warm by the heat of the earth. Therefore, a Dew Pond will only very occasionally drop below the temperature required for dew to condense. There seemed to be little doubt that 'Dew Pond' at least was a misnomer.

Defenders of the faith, however, retorted that Dew Ponds are constructed so as *not* to retain heat at night. As with everything regarding these ponds, no one agrees on very much; many designs are described, and no doubt many existed. In brief, a shallow dish shape is dug, anything from 20 to 200 feet in diameter, often with a raised rim. It must be deeper than the anticipated pond to accommodate the essential lining which will keep it watertight. The lining consisted of clay (where available) or chalk, which was pounded in situ to a fine dust to approximate clay. Lime or soot was sometimes incorporated into the lining material to deter worms. Two additional materials were frequently used: straw, and rocks of a suitable size to lay over the otherwise finished pond to protect the delicate lining from the hoofs of cattle and sheep.

A thick layer of lining clay (9 inches is mentioned) is laid in the hollow. If the layer is to be chalk, some of the loose diggings are returned to the pond and either pounded (puddled) with wooden implements or (and I like this idea) oxen were walked round and round for several hours pulling a heavy, broad-wheeled cart behind them. For some ponds that was the end of it, save that straw would be spread on top to prevent cracking while the pond filled up and/or the above-mentioned stones were laid. Other designs had the straw mixed with the lining material, anticipating fibreglass in its action by holding the lining together. In a twist to the story, and the clever bit spotted by the faithful, some ponds had the straw *under* the lining material or even sandwiched between two layers of lining. The thought here was that the straw insulates the pond from the heat of the ground to allow it to drop in temperature to a point lower than that of the air, permitting condensation to occur.

The straw must remain dry for the lifetime of the pond for it to maintain its insulating properties – possibly for a hundred years. This seems a hopeless ambition, especially considering that the clay and probably the powdered chalk covering it are damp to start with. One investigator dug up part of an old pond to see how the straw was faring and discovered it to be 'moist, completely crushed, and very brittle'. Champions of the dew theory were losing the argument, and things were only going to get worse.

The observations that Dew Ponds can survive a drought seem to carry some weight but are prone to exaggeration. During a three-month drought in 1920, only nine out of thirty-seven Dew Ponds examined in Sussex were recorded to have held on to their water. Another discouragement is that some Dew Ponds may be impostors, being instead deep 'brickholes' left after the excavation of clay deposits for brick-making.

The modern consensus follows Slade, who decided that 'rain ponds' was the appropriate name. He noted that careful observations had found that ponds almost never gained water overnight in the absence of rain. They collect rain, nothing else, and that is the end of it. For the romantically inclined, myself included, this is a great disappointment. However, the modern authorities to whom modern writers refer are as vague as those of earlier times and the most oft-referenced source offers nothing but one paragraph of assertion. Surely there is some hope for us romantics?

Dew aside, there are a few considerations that might protect the Dew Ponds' reputation for being at least a little better at keeping their water than other ponds. It has been suggested, for example, that, being wide and shallow, they form a relatively large catchment area for any rain that does fall, and that rainfall is slightly heavier on top of a hill. It is also possible that they avoid three problems that beset lowland ponds. First, some of the latter, though not all, depend on the water table, so during a drought the table may fall and drain the pond. Second, many lowland ponds are weed-infested, while Dew Ponds are usually kept weed-free to avoid their roots penetrating the clay lining.

This tidiness comes with the crucial added benefit of massively reducing water loss through transpiration by plants. Finally, lowland ponds frequently silt up from material washed down from the surrounding countryside, whereas Dew Ponds have little or no surrounding countryside from which material may be washed.

I write this in high summer and look out every morning from my loft 'office' on the other side of the same Dew Pond hill I could see across the valley from the farm cottage years ago. Nearly always it is veiled in low cloud: that is, fog. In 1919 Edward Alfred Martin wrote a long book detailing his opinions on and experiments with Dew Ponds. His argument was that dew contributed little to its eponymous ponds and that it was mist and fog that did the trick. This seems plausible, as fog consists of very fine water droplets suspended in air. These fall slowly, but still they fall and they do not need to condense as they are already in liquid form.

Unfortunately, the water that Martin argues is provided by fog is insufficient to balance even the loss due to evaporation, and the final defences above do nothing to save the more appealing notions surrounding either dew or fog. It is thus with regret that I accept Slade and the modern consensus. It is Mr Rain that dunnit.

Strip Lynchets

To erie this half-acre holpen hym manye;
Dikeres and delveres digged up the balkes
William Langland, *William's Vision of Piers Plowman*

There is one mystery of the British countryside that I can see from where I am sitting as I write. Visible through the window of my office are a couple of large and long steps in the grass on the side of the hill above my Dorset village. They are Strip Lynchets and, though not particularly impressive examples of their kind, still they evoke contemplation. Who made them, why, when and, most of all, *how*?

Strip Lynchets are simply terraces on hillsides that were once, and sometimes still are, used for growing arable crops. Frequently, they consist of only two or three shallow steps, but sometimes there can be ten or more, marching up a steep hill and anything up to 500 metres long and 30 metres wide. They swirl around valley-sides, vaguely horizontal, following the topography of the hill and generally tapering to nothing or simply fading away at the ends. Most form a fairly gentle stepped incline, maybe 6 degrees, but many are very steep at 20 degrees or more. The steepest on record is 27 degrees to the horizontal, which is a one in two inclination – about half as steep as a domestic staircase.

A little to the east of those I can see through my window there is another, much more impressive, flight of Lynchets. The five treads are a substantial 4–5 metres high, and the entire and rather complex system is 600 metres long, with the longest individual tread stretching for 400 metres. Their average tread width is 15 metres and average inclination 15 degrees. The ends disappear into modern fields which have clearly cut into and flattened the treads. This is a fairly typical Strip Lynchet system, though unusually straight owing to the topography. The system is also typical in that it is very close to the village, a mere 500 metres from the centre. Its nearest sister system, 3.5 kilometres away in the next valley, is also very close to a village.

There are easily a thousand Strip Lynchet systems across Britain. Most, perhaps three-quarters, occur in Wiltshire, Dorset, Somerset and Gloucestershire, mostly on soft Jurassic limestone, the chalk. A substantial cluster can be seen in the Dunstable and Luton area, a few in Northumberland, Yorkshire and Lincolnshire, and a handful scattered elsewhere. Surprisingly, they are rare in Sussex and entirely absent from the Isle of Wight. They are built for the most part in chalk, but also in soft calcareous sand and a handful of other amenable rock types.

The name 'Strip Lynchet' is of modern vintage, coined by O. G. S. Crawford in 1923 to differentiate them from the hillside rows of ancient 'Celtic' fields that have attained a slumped aspect due to soil creep. A 'strip' is a long, narrow field and the more obscure 'lynchet' derives from *hlinc*, a Saxon word for a ridge.

There have been several rather eccentric explanations for the origin of Strip Lynchets over the years. Some have thought them to be natural phenomena such as collapsed faults in the underlying chalk, or even ancient, stranded beaches. Accepting their origin at the hand of man, Ida Gandy, in *A Wiltshire Childhood*, published in 1929, whimsically speculates on their attribution and purpose: ancient Britons who used them for corn, the Romans for vineyards, or perhaps Celtic tribes who

Fairly steep Strip Lynchets in Dorset

once built them as platforms from which they defended against ravening wolves.

The early dates playfully suggested by Gandy have been largely dismissed, though the appealing idea of terraced Roman vineyards could conceivably be true and a Bronze Age provenance for at least some certainly so. The consensus, however, is that most originated in the Middle Ages in response to 'land hunger'. One would imagine that there was always enough land for the relatively small population of those times, and for the most part there was, just. But much of the soil was poor, or if it was good, then it could quickly become poor through over-exploitation. The valley bottoms, being relatively wet, were unsuitable for growing corn and pulses, the staple crops of those times, and were used instead for pasture. It was therefore on the dry, lower slopes that most of the arable farming took place. The tops of the chalk hills, usually a considerable distance from the village, were reserved mostly, if not entirely, for sheep, as were the steep valley sides on which Strip Lynchets are now found.

There are several contenders for situations and periods during which the Middle Ages suffered land hunger, but the period preceding the Black Death is likely to be when the most serious occurred. Food production had increased steadily for a few hundred years before this ultimate calamity, but the birth-rate had increased even more rapidly, resulting in a population that could be sustained only by frequent good harvests. At the beginning of the fourteenth century the medieval warm period was near its end and the climate went into one of its occasional huffs, producing cool, wet summers and early autumns. The wet spring of 1315 made it impossible to plough and to sow corn, and the wet summer ensured that much of what was sown rotted in the ground. The result was the Great Famine of 1315 to 1317. This time frame is slightly misleading, as food supply did not come close to matching demand until 1322, and repeated famines occurred throughout the century. With people desperate to grow food at any cost, the well-drained rough pasture of the hillsides was a clear temptation, albeit a challenging one. Strip Lynchets were created (one way or another) on the hillsides during this period in order to increase the surface area of arable land. Tragically, the land-shortage problem was solved not so much by Strip Lynchets as by the plague.

Practicality is everything in agriculture, so any attempt to impose the practices used on level land will face many problems where the land is steep. When viewed from above, Strip Lynchets look very much like the strip fields that were common everywhere, though they are often much more curved. Furthermore, their lengths are either constrained or greatly extended by the topography. A long field is a great benefit, especially on a slope, as the longer it is, the less frequently the unwieldy arrangements of plough and oxen need to be turned. This is nicely exemplified in the 400-metre or, to use a much more appropriate unit of measurement, 2-furlong (i.e., furrow-length) Strip Lynchet near my home that I mentioned earlier.

If the hillside was marked out as a series of long, more or less horizontal strip fields, and these are then ploughed separately for many years, then the long-term result can be a series of steps, with each field becoming relatively level from side to side. How could this happen?

The medieval plough was a fairly simple affair, cutting a furrow in the soil and turning a width of sod upside down and to the side. The first cut taken on what would eventually become a Strip Lynchet would be two sods' width distance from the lower boundary of the strip. Ploughs of the time generally turned the sod to the right, so the furrow would be made with the hill running down from the right-hand side of the plough. Because the plough had started two sods' width away from the boundary line, it would then turn the sod onto the strip of grass left free next to the boundary line. Once the first run has been made, a dry run is required to take the plough back to the beginning, to avoid turning a line of sods the 'wrong' way, and then the process is repeated. On reaching the last (highest) furrow, a length of bare soil will be left exposed, and a ridge, one sod's depth high, will effectively be formed from the turf above. The result of all this hard work is a ploughed field which is slightly higher than it was before on the lower side and slightly lower on the higher side. Repeat for a century or two and you have a Strip Lynchet. There would have been a little more to it than that – a bit of shoring up of the front edge of the treads, adjustments made to the slope of the riser so that it did not collapse, and the thin ends of each field, unavailable to the plough, dug by hand.

This is, on the whole, almost certainly how most of them were created, but there are questions. The dry run back to the start, essential for the creation of such a terrace, involves an enormous amount of wasted energy. This problem could be resolved with a reversible mouldboard, which can be adjusted for left or right turning, but they were rare in the Middle Ages.

The second problem is that the topsoil will gradually accumulate towards the lower (outside) edge, leaving much of the tread as bare chalk. This immediately invites a third question: how could medieval man have ploughed bare chalk? The only answers to these questions are that, yes, they did make that dry run because there was no alternative, that the vagaries of a messy, periodically fallowed, ploughed field would still allow crops to grow, even if it was mostly on crumbly chalk, and finally, that chalk is easier to plough than one might think because the top several metres are 'rotten', broken down into a friable mass by the action of rain and frost.

One other matter comes to mind: many Strip Lynchets look far too neat to have been formed in this slightly haphazard way. There is no doubt that the process just described was intentional, but some Strip Lynchets, at least, look decidedly planned and built. The shallower (less steep) instances were almost certainly created as outlined, but it would have been very difficult even to begin ploughing what might become a Strip Lynchet on some of the steeper slopes as it is so difficult to run oxen and plough at an angle of 15 to 27 degrees, with 20 degrees being considered the practical limit, and none too easy at that.

The alternative or additional theory is that steep slopes, at least, were deliberately dug. This would have allowed the builders to save the topsoil and re-lay it later. An objection might be that it would have been beyond the inclination or ability of medieval people to complete such a task. But lives depended on it and one only has to take a quick glance at the hill forts built by their predecessors, or at a cathedral, to see what they were capable of. Furthermore, cultivation terraces occur all over the world, so it would not be surprising if the British built their own. Archaeological digs have failed to determine the matter with any great certainty, but one study has given support to the 'built' hypothesis.

In the terminology used for Strip Lynchets, the part of the hill that, by whatever method, has been dug out is called the 'negative lynchet', while the part that has been mounded up above the original surface of the slope is called the 'positive lynchet'. One would then expect the volume dug out to be equal to that mounded up: that is, for the volume of the positive lynchet to equal the hollow volume of the negative lynchet. However, in many instances the positive lynchet is considerably larger than the negative lynchet. In some cases they have been estimated to contain three times the volume of the negative lynchet. The extra material was probably taken from the hill above the lynchet since it was near by and, just as important, it is much easier to move several hundred tons of chalk downhill than up.

The conclusion, then, is that both gradual semi-inadvertent *and* deliberate construction have been employed to make Strip Lynchets. However they were made, Strip Lynchets are magnificent structures, a testament to the ingenuity, skill and hard work of our forebears, and a pleasant thing to see on a day in the countryside – or, for myself, just looking out of the window.

Terracettes

Sheep may safely graze.
Salomon Franck, 1713

Terracettes are ubiquitous wherever there is grazing on steep slopes and will be familiar to everyone who has seen such countryside. Their existence, then, is obvious, as (for the most part) is their origin – from the repeated walking of animals that wander more or less horizontally across a slope, grazing as they go. What is surprising is that they have a name, though the great countryside historian Oliver Rackham refers to them as 'terracets'. He confirms that they are caused by animals as they graze and that they are about 3 feet wide. I am not going to argue with Rackham, but I measured about fifty, and the minimum width was 1.2 metres, about 4 feet, though some were larger; ultimately it will likely depend on the slope. Although they can occur on steep, grazed land almost anywhere, it is on chalk and, to a lesser extent, sandstone that they are most conspicuous.

Individual Terracettes will merge or simply fade away, forming intricate and reticulating patterns when seen from above. That they are not sheep or cattle *highways* is evidenced by more substantial and well-worn tracks at the top and bottom of the slope. Despite the frequent 'application' of manure on the tracks themselves, these have been reported as nutrient-poor. No

Terracettes made by grazing animals

Terracettes formed mostly by soil creep

doubt the 'puddling' they receive from passing animals ensures run-off onto the slope below. They are remarkably persistent features, changing only slowly from decade to decade.

My slight qualification regarding the origin of Terracettes is that in geological circles they are considered to be caused by slippage of soil on steep inclines due to lubrication of the turf with wet clay particles, among other mechanisms. Stock involvement is suggested as a contributing factor. I have seen turf slippage on steep chalk grassland, but the ridges tend to be distant from each other, the whole turf looking like a badly laid tablecloth.

Clearance Cairns

I know nothing more of it; it may be merely a clearance cairn, a spot where stones are laid from ground when preparing for crops.
R. Angus Smith, 'Notes on Stone Circles', 1880

Anyone living in a rocky, rural landscape should have little trouble finding a Clearance Cairn, though recognising it for what it is may prove more difficult. These relics of former times are visible wherever farming has taken place on rocky ground. As gardeners will know, there is an infinite supply of rocks in any given area of soil that float to the surface over a period of only a year. I have several diminutive Clearance Cairns in my small garden; one of them, composed of flint and chert, surrounds my ancient apple tree. I have now despaired at the endless supply of rocks and small stones and have taken to chucking them into next door's garden, along with the snails.

Clearance Cairns (sometimes known as consumption cairns) are piles of rocks removed from a field to facilitate ploughing. Although modern ploughs can tackle almost anything, the

delicate and valuable ploughs of the past could not. These cairns, therefore, are as old as agriculture. In Britain, prehistoric Clearance Cairns are among the most frequent remains of ancient peoples. Some of the rocks in old cairns will show the marks of an ard plough, a sure sign that the rock was found the hard way. Clearance Cairns can be round, oval, linear or an untidy mess; they are, quite simply, random piles of rocks.

The unwanted rocks were piled up at the edges of fields, in woods and sometimes in an intractably rocky field which was sacrificed as a 'cairn field'. Quite often, more often perhaps, the rocks were used to build dry-stone walls – 'consumption walls' – and in this they did at least become an asset. There are some spectacular examples of these on Dartmoor, consisting of enormous boulders superhumanly piled with an admirable artistic sensibility.

Identifying a Clearance Cairn for what it is can be an unreliable endeavour. Many are indistinguishable from burial mounds, and some can be natural formations. In general, a pile of rocks at odds with the rest of the landscape is probably a Clearance Cairn, especially if it is along a field edge. Some may no longer be visible as piles of rocks; having accumulated soil and organic material over the years they can become grassed over, forming a hillock. Others will be consumed within a wood, with trees growing out of them. Those that remain a pile of rocks can nevertheless form excellent habitats for lichens, mosses, plants and invertebrates.

Ridges and Furrows

> *Thou waterest the ridges thereof abundantly: Thou settlest the furrows thereof: Thou makest it soft with showers: Thou blessest the springing thereof.*
>
> Psalm 65:10

Looking at the flat, featureless fields that now grace much of the British countryside, it is difficult to imagine how different they would have appeared in the past. There was a time when many, perhaps most, arable fields consisted of rolling corrugations, with each ridge anything from a metre to 6 or more wide and up to a metre high. These dominant landscape features could have once been found almost anywhere, but most have now been lost to the plough. Nevertheless, many remain, a good number of them as spectacular as the day they were abandoned to pasture. Most of these remnants, however, are revealed only as ripples in the landscape from a light dusting of snow or a partial flood lying in the furrows, or by the sun being sufficiently low to cast shadows. I always keep a lookout as I drive around the countryside and have been lucky on many occasions. But for anyone who is not so lucky, or who sensibly keeps their eyes on the road, there is aerial photography.

Such a bird's-eye view was once almost impossible for anyone without bucketloads of time, access to the libraries that contained the photographs and a generous research grant. Even then, the photographs were poor and incomplete. Now aerial photographs are at the fingertips of anyone with an internet connection. Examples of Ridge and Furrow are most visible on land under permanent pasture and often near a village; indeed some are in people's back gardens. Even former Ridge and Furrow fields that have been repeatedly ploughed can reveal their ancient usage, but the signs are fainter and easily confused with the various linear markings caused by modern agriculture. Early-established Ridge and Furrow fields will sometimes display the reverse 'S' shape – a curve at each end, of which more later.

Ridges and Furrows are remnants of agricultural practices that had largely died out by the early nineteenth century. Those practices varied considerably. The best known and by far the most extensive are called, quite simply, 'Ridge and Furrow'. These were central to the operation of the open field system that stretched throughout most of central England. This encompassed the English lowlands in a (very wiggly) triangle from Harrowgate in Yorkshire to Dorset and back up to Norwich – 'champion country' or 'planned countryside', as it is known. There was also 'wooded country', which is everything else, though even here one type or another of Ridge and Furrow cultivation was practised on suitable soils.

Apart from the Ridges and Furrows required for Water Meadows (see p. 19), there are four main types. One is the aforementioned Ridge and Furrow of open fields; the second is associated with the Scottish system of land management known as 'runrig'; the third is the 'lazy bed'; and the fourth is the most recent, generally known as 'Napoleonic'. All require the ploughing or digging of what are effectively very long parallel ditches and the piling of the soil to form a ridge. Telling them apart is easier than one might imagine as the general form, the width of the ridges and the geographical location usually determine which type you are looking at.

I seem to remember learning (or, at least, being taught) about the open field system at school and will not inflict a detailed account on the reader. For our purposes, it is only the division of land and how it was ploughed that matter. The two, three or four enormous arable fields of a manor were divided up into various holdings. These were ploughed in blocks of Ridges and Furrows within selions. Selions were apportionments of land, long, typically a furlong (660 feet), though it varied a very great deal, and the individual composite Ridges and Furrows were anything from 15 to 60 feet wide.

How Ridges and Furrows come about is simply a matter of ploughing just to the left of the centre line, turning the plough to the right at the end of the run and going back up the other

side of the centre. Continue in this clockwise manner, working outwards until the full width of the ridge has been ploughed. The medieval plough always turned the cut sod to the right, gradually moving the soil towards the centre line. It will clearly not amount to much of a Ridge and Furrow the first time around, but repeat this process fifty times and soil will accumulate to form a broad bank (ridge) and leave a wide ditch (furrow) either side. The process of forming a ridge is known as 'gathering up'.

The above is the standard explanation for how Ridges and Furrows develop. There are, however, some questions: 'why?' is one. The accepted reason is that it improves drainage. Many fields now have subsoil drainage, but not so much in the past. The orientation of the furrows would accommodate good drainage, and ditches were dug and scrupulously maintained to ensure that the furrows did not become permanent linear bogs. It may also be the case that the ridges are warmer than the soil around them.

The second question, an objection, almost, is fairly obvious and was pointed out by the agricultural historian Eric Kerridge in 1951. Once the ridge is the required height, there is no obvious way of stopping it getting higher. Kerridge points out that it would be 80 feet high after a few generations. This is clearly absurd as there is a natural limit of slope to any soil type. Still, there will be an optimum height of ridge for any condition of soil and moisture. Clay requires a high ridge, sandy soil a very low one. As Sir Anthony Fitzherbert explains in his 1523 *Boke of Husbandry*:

> *eyther they be great Lands, as with high ridges and deepe furrowes,*
> *as in all the North parts of this Land, and in some sotherne parts also,*
> *or els flatte and plaine, without ridge or furrow, as in most parts of*
> *Cambridge-shiere: or els in little Lands, no Land containing aboue two*
> *or three furrows, as in Midlesex, Essex, and Hartfordshire.*

The obvious answer to preventing absurdly high ridges would be to 'throw down' by reversing the direction of ploughing. This, however, brings us back to the reverse 'S' mentioned earlier, which occurs on land that was once cultivated using oxen. The serpentine shape of ridges has long been noted and appears to be an artefact resulting from the not inconsiderable problem of turning man, heavy plough and eight oxen in a very small space. The team begins its turn by veering to the left, then sweeping to the right onto and around the headland created for this purpose, before ploughing down the other side of the ridge. Once the reverse 'S' is in place, it is not practicable to go the other way (anticlockwise).

A possible solution is to use a plough with a reversible board (a turn-wrest plough) or an extra plough that turns the sods to the left. Neither of these is entirely convincing as the turn-wrest plough was found chiefly in Kent and there is no evidence of anyone having a left-handed plough. Kerridge explains that it is done by 'slitting', running the plough up and down the top (the 'crown') to make a hollow, while scattering soil left and right. There is a certain amount of tidying up to do with a spade, but overall this will give a rounded shape rather than a pointed one and keep the height to whatever is wanted.

Just when the open field system with its corrugated fields was fading from the landscape, new Ridges and Furrows started to be ploughed on old pasture. These are the Napoleonic fields. The emperor did not have a personal hand in their construction, but during the Napoleonic Wars food shortages occurred, leading local authorities to ask landowners to plough pastureland for arable crops. The ridges are generally much narrower than those of the open field, at around 2 metres wide, compared to the 5 to 20 metres of the open field system. They are still visible in some places: I have seen them on Exmoor and there is a good example of one at York, nicely juxtaposed with an obvious example of open-field Ridge and Furrow.

Runrig, or *Roinn-ruith* in the Gaelic, was a Scottish system that was comparable to and largely contemporaneous with the English open-field system. The chief differences were the approach to how the land was farmed and how it was divided up among tenants. Instead of a person having a permanent set of selions, which could have been scattered all across the various large fields, with runrig rights of cultivation to any set of Ridge and Furrow changed hands periodically. I must note here that, while this is the common understanding, not all academics agree, some considering that most landholdings were permanent. It may, however, be that some within a *clachan* ('hamlet') had permanent tenure and some had just a movable share. In defence of the accepted view, records show apportionments not as numbered plots but as one-quarter or one-sixteenth etc.

Runrig was a form of the ancient infield-outfield system of farming. Runrig systems varied greatly from place to place and changed over time, but the basic idea is as follows. A small number of farmsteads (seldom more than five) making up the *clachan* that worked the land were located around the edge of the infield or 'inbyland'. This area was divided into three parts. These divisions took a three-yearly turn at being fertilised by manure collected from the sheds where cattle overwintered, from rotting roof peat and thatch and anything else vaguely organic that came to hand. Not for nothing was inbyland known as 'muckland'. In this way it was just about possible to grow crops continuously. Oats were sown in two of these years and 'bere' in one. Bere is an ancient variety of barley, a six-row barley, that is cultivated on some of the Scottish islands even today. Beyond this was the outfield. Here the cattle were folded using turf walls in those small parts which had been left fallow for around five years. Oats were grown in a quarter of the outfield and the rest, the bulk, was left to recover natural fertility. The oats from the outfield were usually of inferior quality and yield to that of the fertilised infield. Beyond this was the 'muir', which was used as rough grazing and equivalent to the waste of the English system.

All of the infield was ridged and furrowed, and perhaps some of the outfield too, but the muir was not, being permanent rough grazing. Again, drainage was the main reason for Ridge and Furrow, but in the wet climate of Scotland the ridges were higher, at up to 6 feet, and they were anything up to 20 feet wide, though 10 feet was most common. The furrows were often filled with stones excavated by the plough and thrown there to improve drainage while providing a handy place to dispose of what is always a never-ending supply (see p. 65). These furrows nevertheless produced grass, or at least weeds, which could provide some grazing during fallow periods.

Considering that the idea of a movable tenure (an almost defining peculiarity of runrig) has been brought into question, a great deal has been written about it and, most particularly, the problems it caused. Gilbert Slater, writing in 1907, explained how it worked:

They meet, and having decided upon the portion of land to be put under green crop next year, they divide it into shares according to the number of tenants in the place, and the number of shares in the soil they respectively possess. Thereupon they cast lots, and the share which falls to a tenant he retains for three years. A third of the land under cultivation is thus divided every year. Accordingly the whole cultivated land of the townland undergoes revision every three years. Should a man get a bad share he is allowed to choose his share in the next division. The tenants divide the land into shares of uniform size. For this purpose they use a rod several yards long, and they observe as

much accuracy in measuring their land as a draper in measuring
his cloth.

Farming was extremely difficult on the unproductive soils of parts of Scotland and farming techniques primitive – it took five years for cattle to put on sufficient weight to make them worth eating, and if a farmer got back thrice as much corn or barley as he sowed, he would consider himself doing very well. While the intention of the scheme may have been to spread the misery of the poorer soils evenly, its 'musical chairs' approach to land allocation invariably led to dissension – most probably from the poor devil who feared he would starve to death in the three years he might need to wait for a decent bit of land:

Were there 20 tenants and as many fields, each tenant would think
himself unjustly treated unless he had a proportionate share in each […]
and as they are perpetually crossing each other they must be in a state of
constant quarrelling and bad neighbourhood.

That quarrelling could get very heated indeed, as the following newspaper report from 1739 bears sorry witness:

Donald Smith, alias Macphailow, and James Macpherson, both
Tenants in Dunnachton, 3 Miles from Ruthven, having quarrelled
about a small Bit of Land, as their Possessions lay Runrig, came at last
to Blows; and in the Struggle, Smith pulling out a long sharp pointed
Knife, stabbed Macpherson therewith in the Trunk of the Body, so that
he instantly expired.

This at least is contemporaneous, so it is firm evidence that matters sometimes got out of hand. Movable tenancies or permanent, it was a fairly poor system and was eventually lost to enclosure. An article in the *Scotsman* in 1794, written at a time when runrig had largely been replaced, compared the modern methods very favourably against runrig as practised sixty years earlier. There is a certain amount of 'never had it so good' about this, in that it seeks to show how much worse things were for tenants in the past. Decrying their living conditions, the author writes:

> *They kept their cattle in the same house with themselves, tied to stakes in one end of the houses [...] their food consisted of broth, pottage, oat meal slummery, and greens boiled in water.*

He goes on to say that most of their meat was from animals that had died of starvation or disease. This is bad enough, but even neighbourliness would fail:

> *a single farm was let in runrig among a number of tenants, which caused them to live in a constant state of warfare and animosity.*

Despite its size and once vast extent, most runrig has disappeared to the casual eye, with aerial photographs being the best way to find it. Here it will appear as faint shadow lines in fields. Most has fallen to the plough, with even locations marked on old maps as being runrig showing sparse physical evidence.

A related system, and in some ways the most fascinating, is the lazy bed. This is found scattered on 'difficult' land across Britain, mostly in Scotland, where it is called *fiannegan*, and especially the Western Isles and Shetland. It was also practised in Ireland. George Henry Andrews wrote concisely, helpfully, inaccurately and scathingly in *Modern Husbandry: a Practical and Scientific Treatise on Agriculture*, in 1853:

> *In Ireland a favourite plan of planting potatoes is, on what are properly called, lazy beds; in this case the sets are laid on a piece of ground and the earth taken from trenches cut parallel with the beds spread over them.*
> *The plan is peculiar to Ireland, and it is better that it should remain so.*

The inaccuracy, of course, is that he confines the practice to Ireland. The term 'lazy bed' is unfortunate in that no work-shy person would countenance such a system as a mainstay of life unless he or she was starving. But it was generally confined to areas where no one would normally perform any agricultural activity apart from keeping a hardy breed of sheep, so hunger was probably the stimulus.

The hard work it entailed was daunting because it was necessary to dig by hand in the frequently rocky soil. The implement used was known in Ireland as a *loy*, and a related implement, the *cas-chrom*, was used in Scotland. The latter looks remarkably like a very long, thin leg with a foot attached sporting a distinct heel, the foot at 120 degrees to the leg. The heel of the implement was where the (real) foot of the farmer was placed to do the digging.

As with other Ridges and Furrows, the idea is to pile up soil into a ridge by turning over turfs, but with only two rows of turf involved and with a long turf-width gap left in the middle. After a number of preparatory cuts with a spade, the turning of the thick sods is achieved with the *cas-chrom*, which is driven under the turf to loosen it and then levered over on the heel (of the *cas-chrom*). The chief difference between runrig and its agrarian cousins (apart from it being a manual task) is that soil from the ditches that are formed is dug up and placed in a heap in the above-mentioned gap.

This is the general principle and would work nicely in the back garden should you wish to do something more useful with your lawn, but surviving lazy beds are 3 metres from ditch to ditch, which is more than would be accomplished using the three widths of turf indicated here, so more turfs must have been dug or they cut larger turfs.

Lazy beds were a way of using soils that were difficult to plough, being as much rock as soil, and both peaty and badly drained, described in newspaper reports as 'bogs, wet peat moors and lands too rough to plough'. The advantages of having Ridges and Furrows are obvious in poorly drained soils: 'if the soil be at all damp, plant them [potatoes] in lazy beds: for in keeping them dry depends their safety, in damp lies their destruction.' The soil also benefited from repeated deep digging of the furrow, which threw up relatively nutrient-rich subsoil to the top of the ridge: 'After this trenches are dug on each side of the bed, which usually is from four to six feet wide, and the soil, and often a portion of the subsoil, is dug up (no easy task in hard ground), and equally distributed over the bed to the depth of three or four inches.'

Lazy beds, like runrig, were cultivated continuously, fed by manure. However, seaweed (where available), collected from the shore (especially after a storm), was an invaluable alternative to animal waste. Both seaweed and manure were carried in baskets, always, it seems, by women.

Although lazy beds are best known for growing potatoes, in Scotland at least there was also a crop of corn every year, the potato harvest cleaning the ground in preparation. The potato is a relatively recent addition to the European diet, post-dating the origin of lazy beds when they were used just for corn.

How lazy beds were managed is, as I mention above, not completely clear. The following, albeit from Ireland, may give a clue: 'In the bed system you renew and invigorate the surface on every breaking up, by an addition of the virgin earth from the bottom of the furrows.'

The digging up of potatoes would itself effectively plough the ridges, and I doubt whether 'breaking up' involved flattening the whole thing, just digging it over. It seems very unlikely that the enormous surviving ridges, which must have been bigger still when they were abandoned, were levelled and rebuilt from scratch every year.

One of the most spectacular British landscapes, both from the land and from the sky, is the coast of the Isle of Lewis. From the land the vista is of a green sea of rolling waves. From the 'satellite' view, it depends on how closely you zoom in. From afar, it looks like any other island with a wrinkly coast. Closer, a series of long, closely spaced, straight lines appear, which delineate very thin and very long enclosures of the croft system. Closer still, and underlying these later long fields, is block after block of lazy beds, entwined and nestling with one another.

Storage Pits

And take thou unto thee of all food that is eaten, and thou shalt gather it to thee; and it shall be for food for thee, and for them.
Genesis 6:21

A wander around the countryside will quickly find a hole in the ground. They seem to be everywhere, and it is not easy to know quite how they got there. On a chalk hill near my house there are several shallow holes filled with flint which I am sure are just collapsed natural rabbit burrows. And at a familiar stomping ground in Devon there are some holes in a row which I am reasonably sure are Second World War bomb craters. Chalk pits are another, with OS maps finding them everywhere.

Then there are Iron Age Storage Pits. These are among the

most abundant of Iron Age remains and were used for storing grain and possibly dairy produce. Some are still visible, especially where an archaeological dig has revealed them, but most have been filled in or have filled themselves in and are only readily detectable from aerial photography. The latter technique of finding them and other Iron Age structural remains has been a great success. A recent and extraordinary find was near Winterborne Kingston in Dorset, hiding in a featureless desert of arable fields. Here a previously unsuspected pre-Roman Iron-Age town, nicknamed Duropolis, was discovered. In addition to nearly 150 roundhouses, there were 120 Storage Pits.

Recently I took a day trip to Worlebury Camp, an Iron-Age hill fort near Weston-super-Mare in Somerset. It is moderately famous for its Storage Pits. Almost all of the hill fort is covered in trees, which are busily damaging the archaeological remains, but there is a small cleared area towards the eastern end. It was a little

Storage Pits at Worlebury Camp

overgrown on the summer day I was there, but still the Storage Pits were clearly visible. I counted twenty, but over the whole site ninety-five have been discovered.

Storage Pits are generally circular when viewed from above and often in the shape of a beehive when seen from the side, not that you can see them from the side. Some are just cylindrical. The size varies a great deal from 1.5 to 3.5 metres in diameter and they can be up to 2 metres deep. Some are large enough to have steps leading down. All are carved out of solid rock, something that is easy enough in chalk but difficult in the Carboniferous limestone of Worlebury Hill.

Most archaeological investigation is a matter of digging and surveying, but there is another way – experimentation. Just such an experiment has been taking place at Butser Hill in Hampshire. Here a replica Iron Age village was constructed during the early 1970s and is still there today. The principle is to attempt the replication of presumed Iron Age practices to see if they worked. An example is the idea that thatched roundhouses would always have a hole in the centre of the roof for the smoke. In practice, it was discovered that putting a hole in the roof results in the thatch catching fire from rising embers, whereas without a hole through which it might escape the smoke disperses through the thatch.

They dug out eleven pits of varying size. Some were left unlined, some were lined with clay, some with hazel basketry and one was excavated inside a roundhouse. All were filled with an early variety of barley, and temperature and CO_2 sensors were situated strategically. All were capped with clay in October and opened again in April. This was repeated from 1972 until 1976. The results were encouraging considering the English weather, with only the wet 1975–6 season causing a problem with 28 per cent losses. Even those grains that had become wet and started to germinate were usable, for animal feed at least. Naturally, the pit dug within a roundhouse did particularly well, and one wonders why the Iron Age peoples ever situated them anywhere else.

Pillow Mounds

They forgot the ways of wild rabbits.
Richard Adams, *Watership Down*

Fifteen miles from my West Dorset village is Pilsdon Pen, the second-highest hill in Dorset and the site of many a family walk and picnic. It is an Iron Age hill fort, complete with well-preserved ramparts. As with all hill forts, it has a flat top because, of course, that is where people lived, or at least, where they escaped to when life and limb were threatened by enemies or wolves. Inevitably, this area is covered with the generally inexplicable lumps, bumps and holes that are familiar to anyone who has walked so ancient a landscape.

On Pilsdon Pen there are several that look, quite simply, like graves. However, if graves they are, then they must be those of either giants or men buried end to end and side by side, as the shortest is 10 metres and the longest 40 metres, and from 4 to 6 metres wide. Some odd species of barrow, perhaps? But I am quite used to throwing up my hands at such archaeological details, and for years I thought little about them.

An ever-helpful friend was eventually to provide the answer to this almost forgotten riddle: they are particularly fine specimens of Pillow Mounds. These are artificial rabbit burrows built on the orders of a wealthy landowner for food and fur. In Richard Adams's charming and (fantasy and artistic licence aside) accurate novel *Watership Down*, the company of rabbits looking for a new home encounters a warren that fits the bill nicely for a collection of Pillow Mounds. The inhabitants, kept for the benefit of man, are anthropomorphised as louche philosophers, their spirit gone, their souls and bodies purchased. They have forgotten the ways of wild rabbits.

It is puzzling to think that rabbits ever needed purpose-built accommodation, as they seem all too capable of building it themselves; it is stopping them digging up a field or hedge bank that is the problem. Yet the rabbit is not a native species to Britain but one introduced in the twelfth century. It may well have lived this far north before the last Ice Age, but with the encroaching ice it was eventually restricted to Spain. Evidently too well accustomed to the warmth of the Iberian peninsula, it struggled to survive in twelfth-century Britain and needed a little assistance to make a living here. Incidentally, I rather like the idea that Spain was named after the Phoenician for 'land of the rabbit', though other, more pedestrian etymologies are available for the less romantic.

Norman lords and the Church, both in possession of vast estates and with a line to the monarch who could grant rights of free warren, were able to farm these fashionable delicacies. Free warren removes any risk of being held accountable for

killing a game animal on designated parts of the land. All land was ultimately in the possession of the monarch, and the rights conferred when granting possession to lord or Church were carefully limited. When something new like keeping rabbits was mooted, separate permission was required.

Pillow Mounds are quite common, with perhaps 2,000 still in existence, and were commoner still until fairly recently, when so much land was taken into arable and ploughed. Most are found in south-west England and in Wales. The dimensions of the mounds mentioned above encompass most of the range displayed except they are from 2.5 to 6 metres wide. They are seldom imposing constructions at only three-quarters of a metre high, though they may have collapsed and eroded considerably since their construction. Most Pillow Mounds have a ditch or wide depression all around, from which much of the soil and stone used in construction was taken, a ditch also being helpful in keeping the burrow dry. When on a slope they may follow the contours or run up and down, and are generally scattered throughout a field. Nearly all are long-pillow-shaped, though some are round and there are splendid specimens in Pembrokeshire in the form of a cross.

Some Pillow Mounds found in Wales seem to have been constructed carefully, with slabs of stone used to make runs and chambers. However, John Simpson, head gardener at Wortley Hall, north of Sheffield, writing in 1893, advises a more relaxed approach: first take several pairs of thick turf and lean them together to form a series of inverted V-shapes at intervals around the proposed mound. Each V should be large enough to insert a rabbit. The rest of the mound would be built encompassing these Vs. This will result in a finished mound with little upside-down V-shaped doorways, into which the rabbits are duly inserted. This, he explains, prevents them wandering disconsolately around the fence and saves both time for the warrener and distress to the rabbit. Frankly, I prefer the first method – it would be like building a sandcastle and nearly as much fun.

Where there are rabbits there will be animals who wish to eat them: foxes, badgers, birds of prey and (most troublesome) poachers. Rabbits can also just wander off to found their own burrow or join another one. Pillow Mounds, of which there may be very many in a large field, would then be fenced or guarded by water, and there would be a warrener keeping a lookout from a small hut or dwelling. The taking of rabbits would almost invariably involve ferrets and nets. I have never been ferreting but have watched ferreters in action. It is quite remarkable to see the speed at which rabbits leave their burrow when chased by a ferret, their escape coming to a crashing end in the net and with the push of a practised thumb against their neck.

Pillow Mounds is what these constructions are called now, but they went by a series of related names in the past. The word 'rabbit' was once reserved for a young individual; the name for the adult or the animals en masse was 'coney', a name still applied in North America. 'Coney' is derived from a cognate of 'cunning', a reference to the creature's ability to evade predators. So the names for an 'earth' in which rabbits lived were variations on 'coney' + 'earth': cunningerthe, coningarth, conyger, conigree and the slightly eye-watering cunnery.

In the early days, owning a rabbit warren, complete with its component Pillow Mounds, did more than supply 'exotic' food and valuable fur, it conferred status. It has been speculated that when the silk merchant Sir Baptist Hicks had his Chipping Campden mansion built in 1613, he ensured that the Pillow Mounds were visible to guests from the East Banqueting House. Rabbits were also important to the Church because of their unlikely but official status as 'fish', thus making them acceptable as food on Fridays for most people and for all 180 days of fast for the monasteries.

Eventually rabbits became acclimatised to the vagaries and trials of the British weather, or, to put it more accurately, those with the genetic make-up to survive the weather prevailed, while those whose lineage lacked it died out. Finding both soul and spirit during the seventeenth century, they became thoroughly naturalised. By the nineteenth century the thirteenth-century price of five pennies for a rabbit, equivalent to £12.50 now, became unthinkable. By then it was the food of the poor. But still warrens existed into the twentieth century. My erstwhile home in the village was on a farm that was once just that, a warren. The former tenant farmer there, my old friend Gerald, told me that it was close-wired all round with little doors that let rabbits in, but not out! No Pillow Mound was ever there; as far as I know, it was always open downland. The rabbits are still there, turning the ground a sea of brown fur with bobbing white flashes. They are seldom troubled by shotgun or ferret, but only by foxes and myxomatosis, though the last I suspect is much worse, and also the reason that rabbit pie is now a minority pursuit.

Ha-Has

But unluckily that iron gate, that ha-ha, give me a feeling of restraint and hardship. 'I cannot get out,' as the starling said.
    Jane Austen, *Mansfield Park*

Miss Bertram, the speaker in this Jane Austen quote, truly does not appreciate the purpose of a Ha-Ha. Far from engendering a sense of captivity, they were designed for one of freedom. Nevertheless, our heroine is quite right: Ha-Has are a con, giving

Part of the Ha-Ha at Kingston Lacy, Dorset

the impression of freedom where none exists. Most people will have seen a Ha-Ha – few National Trust properties being without one – but for those who haven't, just think 'grass infinity pool'. Ha-Has are designed to present an open vista with every prospect pleasing, while ensuring that cattle and sheep will not make a mess of the lawn. The name is presumed to reflect the surprise when one sees how the trick is performed: 'Ha Ha!' More sober voices have, however, suggested a contracted and repeated descendant of the Saxon *haga*, the term for an 'enclosure' that is also likely to be the ancestor of 'hedge'. The alternative spellings (and pronunciation) 'Haugh-haugh' and 'Haw-haw' have fallen out of fashion, possibly due to their disagreeable wartime association.

A Ha-Ha is effectively a dry ditch, with the side nearest the house (or whatever vantage point it was designed to serve) generally being higher than the other side. Frequently, the ditch side near the house will be almost vertical and supported by a turf-topped brick or stone wall. Entirely grassed ditches also exist, at one time, and sometimes still, fenced along the bottom. The crucial issue is that the edge of the ditch side, and any wall or fence, should not be visible or, at least, noticeable from the house. Further still from the house there may be a second Ha-Ha, consisting of a broad ditch with a stock-proof wall situated in the middle, running the length of the bottom and effectively dividing the ditch into two: a 'sunk wall'.

Ha-Has became very fashionable with the arrival in the eighteenth century of a more naturalistic and romantic approach to gardens than the formalities that came before. Romantic intentions, however, were sometimes fulfilled with military technology as a Ha-Ha is effectively a defensive ditch. Indeed, many had wooden spikes sticking horizontally out of the supporting wall: a clear case of over-engineering for the sake of show. Ha-Has are thought to be of French origin, though there is an argument dating back a couple of centuries that they were invented by Charles Bridgeman, the landscape gardener at Stowe, where he built the first British Ha-Ha. Originator or not, his contemporary plan of the gardens has a distinctly fortified appearance, and an early drawing shows some truly vicious spikes. Not so romantic after all.

Fairy Rings

[…] you demi-puppets that
By moonshine do the green sour ringlets make,
Whereof the ewe not bites, and you whose pastime
Is to make midnight mushrooms
    William Shakespeare, *The Tempest*

I travel around the countryside every year looking for fungi, perhaps to lead a fungus foray, and it is helpful to plan ahead. If I am destined for pastures new, I will always look online to view the habitat from aerial images. I can usually tell vaguely which species of trees are present, and how much permanent grassland there is. I can also see if the grassland supports mushrooms. While the mushrooms themselves are seldom visible, their rings certainly are, appearing as faint circles and arcs. On the ground they can be seen quite clearly as circles or partial circles of lush, dark grass. Sometimes these dark green rings are encircled by another ring, one of pale-brown dying grass; sometimes mushroom rings are evident only when the mushrooms themselves appear, leaving no trace in the grass itself.

Mushroom rings are always a joy to see and can be found in most fields that have long been left for pasture. Look across a valley and you will probably see several, appearing like giant coffee-cup marks on a green table. They are also common in more urban settings, parks being the obvious place to look for them. I always try for a window seat on aeroplanes as then I can get a good view of the mushroom rings in the grass by the runway. Copenhagen airport, for example, has hundreds of them.

Fairy Rings, for it is these that I see, can be vast and ancient structures, permanent while ever-growing. Among the largest recorded is 800 metres across and perhaps over a thousand years old. Around a hundred British species of fungi are thought to grow in rings, though I suspect it is more. I have seen rings of Parasol Mushrooms sporting seventy or more specimens. One Field Mushroom pasture I used to frequent many years ago was adorned with over a hundred rings.

Rings need be neither circular nor complete. If the soil is irregular physically or chemically, or if another fungus has already staked a claim, then growth in one part of the ring may be faster or slower than elsewhere, or even come to an abrupt stop and die away. One large ring in the above-mentioned pasture looked almost exactly like a map of Australia. Sometimes rings can be almost linear, the remnants of an enormous ring, most of which has died out. Frequently, one will see just an arc of lush grass and maybe its companion mushrooms, nestling among the green stems. If you follow the now imaginary circle around, quite often you will find more mushrooms in another remnant of the same ring. How, though, do mushroom rings form?

At a loss to explain almost any natural phenomenon, people in the past invariably resorted to invoking the supernatural, though such a distinction was moot and seldom made in the past. By a general and almost worldwide consensus, with mushroom rings it was fairies, hence 'Fairy Rings'. They would, it was said, dance around in a circle, scattering fairy dust and causing the various discolorations of grass plus a prompt appearance of the no less magical mushrooms. Nothing good ever came of dancing around in a circle; it is, after all, associated with witches, black magic, the hokey-cokey and so on. And the fairies under consideration were not necessarily *good* fairies, with mischievous elves and even demons also being in the frame. The further consensus, then, was that Fairy Rings are cursed and that straying into one would result in the trespasser being jumped on by giant toads or some such gruesome nonsense.

A few, however, have considered them to be lucky, and in these less superstitious times that is just what they are. Many of the mushrooms found in mushroom rings are edible and, such indulgence aside, show that the ground is productive of something other than just monoculture grass and will probably contain many other fungi, plants and invertebrates; species-rich, in other words.

Needless to say, fairies have no wand in making mushroom rings, but it wasn't until the eighteenth century that any serious scientific thought was given to their true nature and origin. The occasional habit of various animals going round and round in circles during a courtship ritual and producing a fertilising dung while doing so seems a reasonable naturalist explanation, but it is simply wrong.

The most popular theory was electrical in nature. Benjamin Franklin developed an early theory of electricity in the middle of that scientifically productive century. Electricity was the quantum mechanics of its day and, like quantum mechanics, was dragooned into explaining everything. The 'lightning theory' of the origin of Fairy Rings had already been mooted a century before, but now it had a potentially more scientific basis. Erasmus

An enormous partial ring of *Lepista luscina*

Darwin was the chief proponent of this hypothesis and offers a description of how it might work.

Lightning, he presumes, is a column of electricity a few metres across. Since the column is full of electricity it must otherwise be a vacuum. The circular grassy area that is impacted by the lightning is not scorched ('calcined', in his words) because there is no air to allow combustion. The grass at the edge, however, is not within the vacuum and will burn. This causes the brownish ring, releases nutrients to produce the lush, green ring and, thus fertilised, this will attract the growth of mushrooms.

Standing, as we do, on the shoulders of giants (Erasmus Darwin not the least among them), we can see several fatal problems with this hypothesis, the most damning of which is that a lightning bolt is only 2 or 3 centimetres in diameter.

The truth was first suspected and shown by another eighteenth-century notable, the truly brilliant polymath William Withering. A botanist, physician, chemist and geologist, Withering was not a man for wild assertion, but rather a devotee of the only true path to discovery – investigation. It was simply a matter of digging up part of a Fairy Ring and seeing what lay beneath. He discovered, of course, the fine white fibres of the fungal mycelium (the main part of the fungus). Fairy Rings are caused by fungi.

Although it could scarcely be simpler, it is easy to lose oneself in the long description that the process nevertheless requires. A brief summary may prepare you. The underground fungal mycelium grows outwards from a single spot, feeding on its leading edge and dying off as it consumes nutrients. In doing this it naturally forms a mycelial ring. The ring of grass behind the ring of fungal growth is stimulated by released nitrogen from a breakdown of organic matter produced by the fungus as it passes over during its growth.

Here is a more fulsome description. Spores germinate in a single small area of a field. When a spore germinates, it will bud a single, very fine, hyphal strand (under the microscope, spore and strand together look like a frying pan seen from the top). This strand will grow and, all being well, will fuse with other hyphae and share genetic material. The hyphae develop into a mycelium (a mat of entangled hyphae) and grow by feeding on organic debris in the soil. They do this by releasing enzymes from their surface and absorbing the resultant simple organic nutrients. The mycelium can only grow outwards, never inwards, as it depletes its food source as it feeds, leaving an impassable barrier of nutrient-poor soil (for fungi) in its wake. The depleted area will no longer support any fungi, leaving a ring of decaying mycelium, with the living mycelium growing outside. The upshot of this, of course, is that mycelial rings get bigger and bigger. Mushrooms will appear once enough mycelial biomass has been accumulated in the growing ring and once conditions such as season, moisture levels and temperature are suitable.

The lush, green rings of grass that are mostly inside the growing ring of mycelium are green because of the nutrients released from organic matter by a fungus – for the most part ammonia, which will form nitrates. The grass further within the ring is sometimes slightly compromised and grows poorly for a while from the disturbance to soil chemistry.

Rings (with or without their mushrooms present) can appear just as a ring of lush green or as a ring of lush green on the inside plus a ring of brown, dead grass on the outside. This latter type can often display an additional ring, a narrower band of lush green, outside the brown. Some mushrooms do not affect the grass noticeably at all. The lovely and brilliantly coloured Waxcaps (*Hygrocybe* species), in particular, show little signs of changing soil chemistry. It is thought that they grow in the upper layers of soil, feeding on mosses and dead root hairs.

But what about that brownish ring which is seen with some fungi? This is outside the green ring and the site of the feeding and growing mycelium. It is also where (or very near to where) any mushrooms will grow. This brownish ring is seen in only a few ring species. The chief among these is the Fairy Ring Champignon (*Marasmius oreades*), notorious for making a terrible mess of lawns and, most amusing of all, bowling greens. It is a very short mushroom, and if you are short and want to produce and distribute your spores to the four winds, it is a bad idea to grow in long grass. This neat little mushroom takes no prisoners by seriously restricting leaf growth with hydrogen cyanide. Grass can also struggle to grow while the soil is so packed with feeding mycelia.

This is not the whole story. There are two main types of Fairy Ring: 'tethered' and 'untethered', the latter being those already described. These are not attached to anything permanently but can continue to grow until they run out of supporting habitat, such as hitting a gravel footpath or a hedge.

Fairy Rings also grow in woodlands, often forming truly spectacular, dense and symmetrical rings that are frequently complete. I have occasionally seen them neatly, concentrically and almost magically surrounding a tree, a possible explanation being that the fungus and tree are of a similar age. Sometimes, these too can be untethered, living on the dead organic matter in the soil and leaf litter. As with fungi of the field, these are saprotrophs, living only on dead organic matter. But there is another mode of life adopted by fungi: living in a mutualist (mycorrhizal) relationship with a plant. Most often, in the context of Fairy Rings, it will be a tree. It is these fungi that can produce tethered rings.

The hyphae of fungi within the soil are very fine and vast in extent. This enables them to reach parts of the soil that the relatively sparse, short and thick root hairs of a plant cannot. When it comes to mushrooms that live with (usually) trees, specialised mycelial structures encompass and partially penetrate the root hairs and, through this interface, transfer to the tree some of the water and minerals to which the fungus has so ready an access. In exchange, the tree supplies sugars to the fungus.

All mycorrhizal fungi are, effectively, tethered – tethered to their host plant or plants. Not all such fungi form rings, but some do, acquiring nutrients from their host and growing outwards as the mycelium expands its domain, with the fruiting bodies (mushrooms) forming only on the leading edge. Some of these fungi may grow into the grass beyond the woodland where its host lives, but, because they do not feed within the grass, any sign of a ring will show not as a change in the colour of the grass but only as an occasional ring of mushrooms.

When the subject of Fairy Rings comes up in conversation on a mushroom foray or at a cocktail party, most people express despair at the carnage they wreak on their precious lawns. 'How the hell can I get rid of the damn things?' they plead. My answer is that they can't and shouldn't and that I couldn't care less how upset they are.

Scarlet Caterpillar Fungus

> *And all because of the Scarlet Death*
> Jack London, *The Scarlet Plague*

I have a poster, acquired in the early 1980s at a meeting of the British Mycological Society, which once adorned the wall of my living room. It was made up of about twenty photographs of insects and arachnids that had been infected with fungi. It was one of the most gruesome things you could imagine. I loved this poster, but public opinion won the day, and it is now rolled up in a corner of the loft along with the one of Samantha Fox.

Certain groups of arthropods suffer mightily from the attentions of a disparate grouping of fungi that live on them. Flies can be found fixed to windowpanes by a mass of fungal hyphae that have grown from their bodies; other flies will gain a coat woven from mycelial threads. Spiders in the family Linyphiidae can be consumed by a fungus called *Torrubiella albolanata*, which covers the spider's body in a thick woolly coat of hyphae (a stroma), dotted with translucent yellow fruiting bodies. All these arthropods are, mercifully, dead by this stage, but no one, dead or alive, wants a mushroom growing out of them.

Most of these fungal infections are seen only rarely in Britain, though many are just overlooked. *T. albolanata*, for example, has a handful of British records, most of which are around Norwich. This brings a familiar suspicion that it records the distribution of mycologists interested in such fungi rather than the fungi themselves. There is, however, one that I see a couple of times every year: the Scarlet Caterpillar Fungus, *Cordyceps militaris*. *Cordyceps* means 'club-head' and *militaris* means 'bearing arms'.

It is a very odd-looking thing indeed, an orange lollipop sticking out of the ground. But although tiny at 3–4 millimetres in diameter and 50 millimetres tall, the fungus is highly conspicuous because of (and despite its common name) its truly brilliant orange coloration. But there is more to this fungus than can be seen above ground. I found the one pictured on the next page in 2020 while leading a fungus foray and politely asked that I be left in peace for ten minutes while I unearthed it – it is a delicate operation. The stem disappears into the ground, where it is attached to the pupa of an ill-fated moth, probably a hawkmoth, and it is just about possible to see mycelial cords surrounding the pupa. The larva buried itself in the safety of the ground to pupate but became (or was already) infected by the fungus. The fungal hyphae within the larva release enzymes to digest the host and, once there is nothing left to eat, produce their fruiting body. This

orange club – more properly, the fruiting body – is covered with bright orange hemispheres, and it is from these that the spores are produced and ejected. *Cordyceps* species are ascomycetes (a division of the fungi that form their spores inside structures known as 'asci') and thus in the same division of fungi as Cramp Balls (see p. 187) and in the same order as Choke (see p. 98).

The genus *Cordyceps* is in the family Ophiocordycipitaceae. There are a handful more of these to be found growing on and in insect species in Britain, but all are relatively rare. However, there are two large species with appropriately large names, *Tolypocladium ophioglossoides* and *T. longisegmentum*. Until fairly recently these were in the genus *Cordyceps*, but keeping up with name changes is a full-time pursuit and who knows what they will be called next week? Surprisingly, these have no interest in insects, preferring a strictly vegan diet of fungus. Even then they are very particular, parasitising only False Truffles in the genus *Elaphomyces*. False Truffles grow underground in pine and spruce woods, so it is there that these *Cordyceps* are occasionally found.

If you are a health food enthusiast, you may have heard of another *Cordyceps* species, *C. sinensis*. (Actually, it is now *Ophiocordyceps* – see what I mean about names?) It behaves very similarly to *C. militaris*, and parasitises Ghost Moth pupae. The specific epithet means 'from China', and it is in Chinese medicine that it used. In fact, it is found on the Tibetan plateau. I have no up-to-date figures, but it is possible that 100 tons a year are collected. This is not much in terms of potatoes, but an enormous number of *Cordyceps* and worth ten thousand times as much.

I cannot leave the subject of *Cordyceps* without mentioning their most extraordinary property – of some, at least. A few species can change the behaviour of their doomed host for their own dark ends. *Ophiocordyceps unilateralis* is a tropical species (so you won't find one in Rutland) which, once the infection takes hold, compels its ant host to leave its nest for somewhere damper and more suited to the fungus. The ant then attaches itself by its jaws to the leaf vein of a plant, 10 inches above the ground. And it waits. After a few days the fungus will have consumed most of the soft interior of the now-deceased ant. With sufficient fungal material now created within the host, the fungus fruits by growing a fruiting body out of the head of the ant. This ant is known, unsurprisingly, as the Zombie Ant.

Tropical rainforests are far away from the tame and mild British Isles, but here too insects are forced into fatal slavery. The familiar housefly, *Musca domestica*, has its own fungal parasite, *Entomophthora muscae*. Within five days of infection the fly will, under the influence of the fungus, search for a high vantage point, land there and die. Before it dies, it will have spread out its legs; its wings and abdomen will be pointed uncomfortably upwards and its proboscis extended. Conidiophores (structures that create asexual spores known as conidia) form over most of the softer parts of the fly, their released spores unimpaired because of the open posture of the unfortunate fly.

Choke

Great fleas have little fleas upon their backs to bite 'em,
And little fleas have lesser fleas, and so ad infinitum.
And the great fleas themselves, in turn, have greater fleas to go on;
While these again have greater still, and greater still, and so on.
　　Augustus De Morgan, 'Siphonaptera'

It takes a rather obsessive observer even to notice Choke. It is something seen out of the corner of the eye as a single or a collection of grass stems that just look rather odd, and even then it is seen but rarely. Nevertheless, the diligent naturalist is certain to find one sooner or later, though I have never encountered an entire field of them, as was reported by the English mycologist and adventurer George Massee, in his *British Fungi, with a Chapter on Lichens*. Despite its elusive nature, Choke is one of the commonest fungi on the planet, infecting about a third of British grass species (all in the botanical family the Pooideae), and with a large percentage of grass plants in any field supporting the fungus. This conundrum, at least, is easily explained: most of the time it is invisibly *inside* the plant, making its existence evident only when and if it produces its reproductive parts: the white then orange collar around the grass stem that we see as Choke.

These reproductive parts first appear as that white sheath, 30–50 millimetres long, on the grass stem. This sheath then (usually) thickens and turns yellow or orange with the developing spore factories known as ascophores. A simple fungal parasite of grass, one might think. But not so: Choke has much better stories to tell.

All tales in the living world exhibit multilayered complexity. Sadly, much of this is, necessarily (or not), hidden in magazine articles and nature programmes, which can often tend to the superficial. The story of Choke, complicated as it may seem, is little different in the complexity that attends most of these tales, and forms a compelling example of the dazzling intricacies to be found in something as unprepossessing as a bit of discoloured grass. There is a lot to take in, and my description is scattered with technical words and Latin names which will not be familiar to everyone, though most technical words are explained.

The Latin name for the best known of the forty or so species of Choke is *Epichloë typhina*, meaning 'upon young shoots' plus 'like a bulrush', together describing the brightly coloured, bulrush-like club that forms on and surrounds the middle of stems (tillers)

of grasses. There is another genus, *Neotyphodium*, of which more later. The Chokes are in the division Ascomycota, and therefore related to morels, truffles and yeasts. One of its closest relatives is the notoriously poisonous Ergot, *Claviceps purpurea*, a fungus similarly hidden most of the time and visible only as a 'sclerotium', a black structure, looking like a large grain of rice, sometimes seen on grass in place of the seeds.

The 'invisible' component of all Choke species is a network of fine fungal fibres known as hyphae. These can be found in nearly every part of the grassy host except the roots. They do not penetrate any cells but live instead between them, running a wiggly line throughout stem, leaf and seed, seldom branching and living on the thin gruel of inter-cellular nutrients.

Normally, hyphae grow only from their tip (apex), while grass stems grow both from the apex and through the enlargement and elongation of established cells throughout the stem. Hyphae are extremely fragile and would quickly fracture within a growing grass stem, so Choke has developed a method (unique in the fungal world) of growing throughout its length, keeping pace with the stem. This is no mean trick but one that depends on a long series of unlikely biochemical changes. The reproductive strategies employed by Choke are no less amazing.

There are at least three broad approaches open to a fungus as a means of reproduction: vegetative (like the runners you find on strawberry plants), asexually produced spores and sexually produced spores. Not leaving anything to chance, Choke goes for all three, the strategies or combinations of strategies varying from species to species and not notably consistent even then.

It was not until 1992 that it was realised that there were many more species than just *Epichloë typhina*. Molecular studies, mating-compatibility experiments and host-specificity observations (which grass it grows with) revealed at least forty. Many were retained in the genus *Epichloë*, but there are some 'Chokes' that distinguish themselves by never producing sexually reproductive structures (the white and orange collars seen on stems) and which

Mature *Epichloë* fruiting body

are hybrid descendants of *Epichloë* species. These were placed in a new and rather awkward genus, *Neotyphodium*, until it fell out of favour in 2014, and I use it here only for convenience. Whatever one calls them, they are common, but because they lack any sexual apparatus (collars on the stems) they require a microscope to see them.

Neotyphodium species reproduce (partly) by the simple method of growing their hyphae into grass seeds and continuing to grow within the new plants that germinate from these infected seeds. This form of vegetative reproduction is a dicey business, with mutations accumulating over the generations, though *Neotyphodium* species do have a few tricks to avoid this fate, notably in possessing a great deal of spare genetic material accumulated from various strains during the hybridisations with which they originated. This technique of infecting seeds is remarkably successful, with most of the seeds in any one plant being infected yet still fertile, and producing a new, infected plant.

Not content with this, these asexual species, living as they do within their host, have been found to produce conidia (a type of asexual spore) on the external surfaces of grasses (e.g., leaves), accompanied and supported by barely visible wisps of mycelium (bundles of hyphae). These conidia, it is believed, may be capable of infecting other grass plants, perhaps through direct contact with a neighbour or using water, wind or insects as a vector. It has also been speculated that in colonising the surface of a plant they fill an ecological niche that might otherwise be claimed by a competitor fungus.

Exactly the same techniques are also found in most, but not all, the sexual *Epichloë* species, some of them forgoing surface conidia, some forgoing the infection of seeds. Sexual reproduction in these species is, as you will by now expect, not exactly straightforward. Sexual reproduction begins with that thin white sheath (stroma; plural: stromata), that surrounds part of the stem. The stroma forms spores, which will be dispersed and may well go on to form a sexual union on another stroma on another grass plant, though it may directly form an asexual, clonal line in a new host plant as with the conidia mentioned earlier. Dispersal is most often courtesy of a fly.

Employing the evident allure of a cocktail of unpronounceable organic chemicals, the stroma attracts the attention of flies from the genus *Botanophila*. These are rather unassuming creatures, notable only for their large red eyes. The fly feeds on the spores and the stroma on which the spores grow, then flies off to find another infected grass. On landing for a second snack, and under the influence of behaviour-altering chemicals in the consumed fungus, it (there is no polite way of putting this) defecates all over the visited stroma by wiggling its rear end in a zigzag fashion to ensure good coverage. The previously ingested spores are thus placed in close proximity to the female reproductive bodies, known as ascogonia, and mating takes place. Once this inadvertent fertilisation process of Choke is complete, the fly lays its own eggs. These hatch and feed on the fungus. When the larvae are of sufficient size they usually drop into the soil, where they pupate. The ascophores propel their spores into the air by the thousands, and they float about, 'hoping' to land on a suitable grass and germinate.

There are hints that another organism is involved in this already baroque process of sexual reproduction: a bacterium. More precisely, it is the bacterial sexual parasite *Wolbachia*, which infects the flies. 'Sexual parasite' brings many things to mind, none of them good, but with *Wolbachia* it is a disruption of the reproductive process in its hosts. This can take many forms, but in this case it makes the eggs non-viable.

One of the problems facing such mutualisms (symbioses where every one of the involved organisms benefits) is their stability. If one species gets the upper hand, the entire relationship can collapse, and all the involved species suffer. It has been hypothesised that the sexual parasite *Wolbachia* is endemic to *Botanophila* but that, even so, the fly can still produce fertile eggs because it is administered an antimicrobial by the fungus. This makes sense, since the fungus's own reproduction depends so perilously on that of the fly. However, the relationship appears to be more subtle than this: the antimicrobial may be fed as a matter of course to the fly while it is behaving itself, but is withheld if it becomes too greedy, preventing the overgrazing of fungal tissues by larvae. Checks and balances.

Thus, we have three species getting on well enough, but what of the grass, which seems to be doing all the heavy lifting in these relationships? It does not look good. Any stem on which the reproductive bodies of *Epichloë* form will be inhibited by the fungus from producing flowers. This is a great imposition on its generous host. Quite why the grass tolerates this, or the burden of carrying and feeding the fungus within its tissues and all the reproductive goings-on without, may seem a mystery, especially as many plants have developed effective methods of combating such infections. Enter the story of Kentucky 31 and the Fescue War.

During the terrible droughts that befell parts of the US in the 1930s, the pasture on which so many cattle depended dried to desert. Many grasses can tolerate remarkably low rainfall, but conditions in those troubled times killed everything. Well, nearly everything.

In Menifee County, Kentucky, during the intense drought of 1931, Professor E. N. Fergus from the University of Kentucky found himself judging a sorghum syrup competition. As he was performing his sticky duties, he heard from a circuit court clerk about some grass that was seriously bucking the trend by remaining lush and green while all about was dry and brown. He went to see a W. M. Suiter, on whose blessed soil it was growing, covering an entire hillside. He took away some seeds and began trials. The story has it that the original seed stock travelled from its native Europe to North America in the mid-nineteenth century as packing straw for some fine china.

The grass, Tall Fescue, *Festuca arundinacea*, proved to be drought-tolerant, vigorous, high in seed and tiller production (lots of shoots) and resistant to soil nematodes and insects. It was discovered that the particular lineage of Tall Fescue was infected by a previously unrecognised species of Choke, *Neotyphodium coenophialum*, which through chemical means conferred all of these benefits, nicely explaining why the grass tolerates its demanding house guest. More than this, the gifts bestowed by the fungus were the answer to a million agrarian prayers.

Production of seed, known of course as Kentucky 31, was set about with great enthusiasm, and soon thousands of acres of pasture were successfully sown with the infected Tall Fescue. However, after a very good start, farmers began complaining that their cattle were suffering strange and sometimes fatal diseases.

N. coenophialum's various tricks are accomplished with a small biological arsenal that includes three alkaloids. Peramine and loline respectively deter and kill insects, while the third, ergovaline (also found in Ergot), deters and sometimes kills grazing animals. The symptoms of ergovaline poisoning are distressing: 'Fescue foot', a localised gangrene caused by the vasoconstricting effects of the alkaloid; 'Fescue toxicosis', which is a failure to put on weight, along with reduced fertility, rough coat and an intolerance to bright light; and 'Summer slump', which is severe overheating and a tendency to stand in the shade or in water.

These effects were, and often still are, catastrophic for grazing animals and those that depend on them for their livelihood. When these problems came to light, there was a prolonged and heated argument between the University of Kentucky's academics – some accepted that there was a serious issue, and some argued that there was not. Academic disagreements can sometimes be vicious, and they were particularly so in this case, which was subsequently dubbed the Fescue War.

Fortunately, it has been possible through breeding and, more latterly, genetic techniques to remove the offending organic compounds while retaining those that are beneficial. In wild populations, of course, things go on as normal, the infected grasses benefiting from reduced predation, increased vigour and resistance to drought. Well worth all those seeming freeloaders.

The Wood

Coppices

Where the folding of sheep is the great source of manure and of crops of corn, rods and hurdles will be the object of his coppice; and here he will want Hazel.

William Cobbett, *The Woodlands: or, A Treatise*

At the bottom of the dry chalk valley near my old home in my West Dorset parish lies a small wood. In truth it is two woods, contiguous but with different names: one is called Parson's Coppice and the other is named after the village, suggesting that it had once been held in common. Much of the slope that rises from the valley floor is of oak, ash and (formerly) elm. The flat areas at the bottom and the top are nearly all of hazel. These hazel stands are magnificent and enormous at nearly 1.5 metres across at the base, each base sporting their customary scores of separate stems. And they are unnatural.

The British countryside is a construct, built by the hand of man over thousands of years. While we accept that truly wild places are rare indeed, we still see our tamed landscape of fields and woods as a thing of beauty. But many of the fields are wastelands of monoculture and the woods recent plantations of introduced trees. Man's interactions with the natural world have not always ended well for our fellow species, but some have been remarkably beneficial to all concerned. Meadows are iconic in British conservation, as is chalk downland. Coppices too must take their place at this noble table. All can harbour hundreds of species of plant, animal and fungus, yet they are made by man. And they are managed.

Much of our native-species woodland is neglected and unmanaged, or at least managed badly. I am sometimes asked to visit someone's wood to identify any of the fungi found there. It can be a depressing exercise. The wood will almost invariably be overgrown with ivy, brambles or pendulous sedge and clogged with plants, such as the dreaded Rhododendron, that have invaded from a nearby garden. They will also be full of unsuitable trees (spindly sycamores being the worst) and with too many trees of almost any sort. Few fungi grow in such conditions and little of anything else. Their wood is unmanaged.

Tended by those whose livelihood depended on them, Coppices were always scrupulously cared for. It is fashionable for those who wish to serve the natural world to encourage us merely to leave nature to itself, but this is a very long-term strategy in most cases and quite possibly a disaster if certain elements, such as

grazing animals and top predators, are missing. In the managed combination of stability and change required for a Coppice, however, things work out perfectly.

Like the wood banks that surround them, Coppices are not notably mysterious – most people will know one when they see one, but not everyone is familiar with how they work and certainly not everyone knows what ecological triumphs they can be. Coppices have been around for a very long time. In Neolithic times the walkways across boggy land were made from coppiced material, and the technique of coppicing is known to have been used by the Romans. During much of the Middle Ages coppiced woodland was by far the most common, supplying fuel, building material for homes and lengths of split stem for hurdle-making and tools, as well as having scores of other uses. It was not until the nineteenth century that Coppices began to disappear, their use as fuel replaced by coal. Since the early twentieth century they have declined much further, with 90 per cent lost. They are now enjoying a small revival, owing in part to the many people who have taken to such things as hurdle-making and traditional hedge-laying, and to conservation bodies that have seen their merits.

A Coppice is an area of land divided into panels (small areas of land) and planted with trees. These are cut back to ground level while still young, then left for a period of years before being cut to ground level again, when they will regrow, producing even more stems. The ground-level part after cutting is known as a 'stool'. When repeated for a few cycles, each stool will produce scores of shoots. The tree species chosen will depend on what use is required of it and soil conditions. Hazel is the most familiar, but willow and sweet chestnut are also common. Ash, oak, alder, field maple, small-leaved lime and wych elm have also been employed. The more substantial wood species will have a correspondingly longer cutting cycle. Stands are always cut in the winter, while the tree is dormant; when cut at other times, the sappy wood it produces does not handle well and is more prone to decay.

Once established, each panel will be cut when the diameter of the shoots has grown to the required size. This may be very small if it is for hurdles (movable fences, woven from hazel or willow), thicker if it is for tool handles and larger still if for structural material or charcoal, though any single stand may contain timber of varying-diameter stems. This means that in a working Coppice some panels will be shaded by a dense tree canopy, a few will be in partial shade and others in almost full sunlight – a cyclical ecosystem rather than a successional one. In English, this means that the normal sequence of developmental stages – grassland, scrub, immature forest, climax forest – is broken at the immature forest stage. This allows shade-loving plants to thrive for a while, then die back to be replaced by sun-loving plants. Some of the latter are quite rare, such as Narrow-Lipped Helleborine, *Epipactis leptochila*, and Narrow-Leaved Bitter-Cress, *Cardamine impatiens*. Most of our uncoppiced forests develop a full and permanent canopy and are dominated by vast, dull stands of wild garlic or dog's mercury, or carpeted with ivy.

It gets better still, as there is variation in the cycles used, with short cycles of perhaps seven years for one stand and up to thirty years for others. Old hazels, for example, are ecologically interesting because they can host many species, such as Toothwort (see p. 148–50). Also, most traditional Coppices contained standard (full-size) trees, which were felled when large enough to provide baulks of timber. This allowed the establishment of species requiring a home on the bark and the time to grow, such as mosses, lichens, liverworts and fungi. Such Coppices are known as 'coppices with standards', while Coppices without standards are known as 'simple coppices', the latter being much more common today. The open habitat afforded by coppicing is also of importance to species of butterfly and bird.

In the past it was common to graze or forage Coppices, usually during the winter, when the stock will do less damage. Nevertheless, a certain amount of fencing will be needed to keep animals away from young shoots growing from the stools. Grazing and foraging in woodland are time-honoured but little seen these days. My friend Oliver owns a marvellous cider orchard in which he allows his pigs to eat the otherwise wasted fallen apples, so such practices have not died out completely. Similarly, Commander Eyre, my one-time landlord for the farm cottage I used to rent on the land that includes the two Coppices, would leave the gate open to his breeder herd of cattle. When the land was bought by the Nature Conservancy Council back in the 1980s, I asked if they would allow grazing in the Coppice. They were very dismissive of the idea, and no grazing has taken place since. The cattle made a muddy mess, but they did keep clear an open area in the middle that would otherwise have been claimed by brambles. This is now covered in a carpet of ferns, and I cannot say that the wood is quite what it once was.

Wood Banks

> '*But, by Jove, he's away!*'... *springing into the field from the high wood bank.*
>> Knightley William Horlock, *The Science of Foxhunting and Management of the Kennel*, 1868

Why would anyone go to all the effort of digging a ditch all the way around a hundred-acre wood, laboriously piling up the soil to make a bank, a 'Wood Bank'? The short and trite answer is that they had no choice: someone told them to. The more sensible answer is much the same as for a hedge or a stone wall – to define a boundary and to exclude (or include) grazing animals.

In a book of mysteries Wood Banks are not exactly an imponderable puzzle. They are banks that have been built around, and sometimes within, a wood. They are very similar to hedge banks, and just because a bank is alongside a wood it does not necessarily mean it is a Wood Bank – the hedge and bank could have existed before the wood. Like hedge banks, Wood Banks consist of a bank and a ditch, the ditch being on the field side, not that of the wood. Both bank and ditch will belong to the owner of the wood, not the next-door neighbour. Medieval Wood Banks were often heroic constructions at anything from 6 to 12 metres wide, if one includes the ditch, so these at least are easy enough to guess at. Later Wood Banks tend to be much smaller, at up to 3.5 metres, which is still more than a hedge bank, and can be wider still if they double as an estate boundary. For any more certainty, it is necessary to look for some other identifier.

A fairly typical Wood Bank

One clue, which is, frankly, both unreliable and faint, is that the slope of the bank on the ditch side tends to be shallower than that on the wood side. Quite why, I do not know, though I have a few speculations with which I will not tire you, but merely offer the rather obvious observation that it was probably just how they ended up, either as a by-product of construction or from some subsequent physical factor that affected one side more than the other. Many old woods are the relics of assarting, where woodland is untidily left over from encroachment by agriculture. This results in a noticeable lack of straight boundaries, and it is a frequent characteristic of Wood Banks that they tend to follow a more sinuous path than hedge banks. Much of the above is moot anyway, as Wood Banks were often topped with a hedge and also pollards (see p. 117), especially if they marked the edge of an estate.

Most woodland in medieval England (and elsewhere) was coppiced (see p. 109), hence the strong association of Wood Banks with older woodlands. Then, as now, animals would need to be kept away from the recently cut stools, at least until the new shoots were established. Unfortunately, coppicing happens within a wood in a progressive fashion, with some stands ready for cutting, some not quite ready and some newly growing from their stools. So somewhere in the wood there would probably be succulent young shoots on which animals could graze. Local fencing in the form of hurdles or a dead hedge made from the brash (mass of small branches) left over from cutting could be used. So important was this for the well-being of the wood that a clause to this effect would be included in leases. Here is one from the late eighteenth century:

> … *also upon every fall which shall be made on the wood or coppice*
> *lands of or on the said demised premises, to fence and copse up the same,*

to preserve the future sprigs therein from the bite of sheep and cattle and
all other trespass …

In medieval times and beyond, woodland was effectively leased by the lord or a priory to another person for coppicing purposes, and they would fence and hedge as necessary. Unfortunately, the wood might also be leased for foraging (usually by pigs) *at the same time* and to someone else. Anyone trying to coppice in a foraged area would have spent a great deal of time fencing off panels, and it is thought that they may have resorted to the more difficult process of pollarding their stands instead.

While internal wattles or dead hedges are understandable, internal Wood Banks need an explanation. Sometimes there would be two people leasing different parts of the wood, and something more substantial would be required to prevent encroachment than an all too movable fence – a bank and ditch. Borders of any flavour are notorious for causing strife, so occasionally a lessee might consider that the bank had been put in the wrong place and would feel seriously aggrieved if the situation was not remedied.

In the twelfth century just such a conflict occurred between the town of Melrose and the nearby Kelso Abbey. The trouble had been rumbling on for years, with Melrose resorting to moving the offending bank and taking possession of what they thought to be something they were paying for. A remedy to the dispute seemed to come in the form of a visiting papal legate, John de Salerno. He listened carefully to both sides for a few weeks, accepted gifts of horses and gold from both and promised both that all would be well. Then he rode away, leaving the towns no closer to remedy, though, with all those horses, one cannot say that he was particularly burdened by his ill-gotten gold.

Pollards

It is odd that so rural a practice as pollarding should have given up its country ways and moved to the big city. While there are many thousands of pollarded trees scattered about the British countryside, few are still tended, few are still pollarded. If you wish to see actively pollarded trees, you must search out an urban avenue, though they are often cut much higher.

In my local town of Dorchester there are rows of tall lime trees not far from the market. They are elegant and very burry (see p. 151) and look lovely in the late spring with their bright and shiny pale green leaves. But being street trees, they have very little room to manoeuvre – households would soon have leaves pressing against their windows, and the two rows of an avenue

A splendid veteran ash pollard, long left untended

would meet in the middle to form a dark tunnel. So, every five years or so they are pollarded, though this is much more often than for their country cousins.

The young growth is cut back to the various 'heads' that have accumulated over the years from repeated cutting. In the winter, the trees take on a sculptural appearance with upward-arching branches, the terminating heads bearing a mass of young shoots. They look as though they might, just, be able to walk. With such a typically 'burry' species, the shoots and leaves of pollarded lime will grow almost anywhere from the tree, giving it a very bushy appearance. When it comes to the five-yearly cut, the result can be quite shocking. Stripped of all its shoots, it looks like a bad haircut. The consolation is (as I once inadvisably told an old girlfriend), 'Well, at least it will grow back.'

Despite its new-found fondness for the bright lights, the roots of pollarding are entirely rural. It is known from Anglo-Saxon times onwards in Britain, and it may have once been the case that most British trees in southern Britain were pollarded. Pollards are found on wood edges and in hedgerows, but it is in wood pasture that they once found their commonest home. Wood pasture is a beguiling prospect: widely spaced trees in otherwise open and permanent pasture. The birds would be singing, butterflies flapping, and the calls of cattle and sheep could be heard. With oak and beech, pigs would snuffle for acorns and mast. An entrancing idyll. How extensive this system of farming was is not known with any certainty, but it is likely that it was adopted throughout much of lowland Britain. It was primarily an Anglo-Saxon practice, slowly giving way to separate pasture and to more extensive coppicing.

The practical appeal is fairly obvious: wood (for fuel, posts etc.) and animals could be produced at the same time and in the same place, but there are two problems. The supply of wood is highly irregular with standard trees, feast and famine being the rule, and trees grow sideways as much as possible in most circumstances, and the circumstance in the case of wood pasture is for a broad spread until the pasture is too shaded to grow. Pollarding was the solution to both these confounders.

The size of a pollarded crown is very much smaller than that of the same tree species growing in what is effectively open land, and it is kept that way by repeated lopping over the centuries. The lopping solves the wood supply problem too, as once the wood pasture is established it is continuous. There was an added advantage to Pollards: they provided fodder in the form of young leaves and thin branches. This practice goes back to prehistory but had largely died out as a wide-scale practice in Britain by the twentieth century.

Pollarding is the systematic and periodic pruning of the top part of a tree trunk at a height of 2 to 3 metres. This encourages the growth of multiple shoots while ensuring that most of them are beyond the teeth of grazing animals. These shoots are cut close to the rounded 'head' or heads, but not too close as the growth occurs below the cut. I can do no better in further explanation than to refer you to our earliest adviser on this subject, Sir Anthony Fitzherbert, writing in his *Boke of Husbandry*. I won't spoil it for you with any explanation, except to note that *slaue* is 'strip' and possibly a misspelling of *flaue* ('flay').

> *Lette hymme begynne at the nethermoste boughe fyrste and with a*
> *lyghte axe for an hande to cut the boughe on bothe sydes, a fote or*
> *two foote from the bodye of the tree. And specially cut it more on the*
> *nether syde, than on the ouer side soo that the boughe fall not streyght*
> *downe, but turne on the syde, and than it shall not slaue nor breke*
> *no barke. And euery boughe wil haue a newe heed, and beare moche*
> *more woode; and by thy wyll without thou must nedes do it, crop not*

*thy tree, nor specyallye heed hym, whan the wynde standeth in the
northe, or in the eest. And beware, that thou croppe hym not, nor heed
hym (specially) in sappe-tyme , for than wyll he dye within fewe yeres
after, if it be an oak.*

Fitzherbert urges the reader not to cut during 'sappe-tyme'.
This is now understood as being any month except January and
February. The frequent newspaper reports of protests when the
local council (Dorchester Town Council being one of them!)
has waited until April to pollard their street trees indicate that
Fitzherbert still has many readers.

It is customary not to cut everything, as the tree will fail
to produce sufficient new leaf to feed the tree. The poet and
farmer Thomas Tusser, writing in 1573, agrees: 'One bough stay
unlopped to cherish the sap.' Depending on the size of wood
needed and the species of tree in question, the cutting cycle is
between fifteen and thirty years. (Urban trees are pollarded
to keep them a manageable size, not for their timber, thus
demanding a much shorter cycle.) It seems likely, however, that
more than one branch would have been left to cherish the sap,
with just a few of the branches cut each year as they reached the
required size: a woody 'cut and come again' salad.

Cutting too close has been a problem for those tasked with
resurrecting ancient Pollards that have been left uncut for a
couple of centuries. This has been tried in several of the (sadly
few) places where these trees abound, such as Burnham Beeches,
Hatfield Forest and the New Forest. In the 1950s it was decided
to tackle 100 ancient beech Pollards at Burnham Beeches, and
the services of a London tree surgeon were sought. The trees
were duly lopped and pruned, but, the surgeon evidently not
understanding the finer points of pollarding ancient trees (though
in fairness no one did at the time), the cuts were too deep and the

High-cut pollarded lime trees in Dorchester, Dorset

resultant fatality rate was nearly 100 per cent. Techniques have improved since then, notably to always keep some branches in place, as farmer Tusser had so wisely advised.

Why one would wish to go to all the trouble of saving trees that may well be 400 years old is a question that most of continental Europe would find difficult to answer. However, the British have always shown a peculiar fondness for ancient trees, with more surviving in Britain than in the whole of continental Europe, except Greece, where venerable olives still cling to life. More than this, old trees become old habitats, supporting organisms that take a great deal of time to establish, such as lichens, mosses and liverworts. Some lichens won't go near a tree until it is truly ancient. Fungi too can take a long time to establish a home. One of the finest places for rare fungi of ancient woodland is Windsor Great Park and another is the New Forest, both places where ancient Pollards live on.

The impressive longevity of pollarded trees is well known. They have found the magic formula of being both old and young at the same time. Being continuously trimmed, they are kept permanently at the juvenile stage of growth with all the vigour that entails. However, the bolling (trunk) of the tree, like that of any old tree, continues to grow and hosts slow-growing brown-rot fungi. These hollow out the trunk, releasing nutrients back into the soil and reducing the tree's overall weight. They also have a helpfully small wind profile for their considerable age. Oliver Rackham points out that the great storm of 1987 was informative in this matter, as the Pollards remained stalwart while standard trees lay in devastation all around.

While eminently practical and prolonging the life of a tree rather than shortening it, pollarding has always been a brutal business. With wood pasture falling out of favour and vocal agriculturalists decrying the existence of *any* large tree in the hedgerow (they sapped the strength of the soil and shaded the crops), pollarding became unfashionable. Its agricultural critics certainly took no prisoners. John Leonard Knapp, writing in *The Journal of a Naturalist* in 1829, makes his position clear: 'There are not many of our rural practices, that deserve more the disapprobation of the landed proprietor than that of pollarding trees.' He goes on for a couple of pages in a similar vein, even evoking a fellow grumpy old man, the Teacher of Ecclesiastes: 'It is an evil under the sun, and common among men.'

The late eighteenth and early nineteenth centuries saw wood pastures abandoned and the land put to a better, or at least other, use. Now we revere and cherish those Pollards that remain.

There is still some active pollarding in the countryside. Willows, for example, are routinely pollarded, and historically important sites such as Hatfield Forest, where wood pasture has survived, have sought to sustain them with new plantings. And there are still all those towny Pollards. Long may they live.

Hollow-Ways

[…] a green thought in a green shade.
Andrew Marvell, 'The Garden'

Some mysteries of the countryside can be revealed only by close observation and careful consideration. Others, however, are perfectly obvious if only we take a moment to think. Years ago I wondered at the enormous effort taken by people of the past to dig out deep paths and trackways to make them (reasonably) level. Thousands of tons of soil and rock seem to have been removed for very little gain; trackways are better if level, but it is not an essential requirement as it is with railways.

Hollow-Ways occur throughout Britain, though only where there is a relatively soft bedrock, such as sandstone or chalk; there is no point looking for one on the granite of Dartmoor, for example. They are familiar to anyone who drives the country roads of Britain where ancient trackways have been tarmacked over. Suddenly, on a bright day, there is a plunge into darkness, the road acquiring an unexpected wall on either side and a cathedral arch of trees above. A Hollow-Way is only visible from the inside; indeed, it is possible to look across a large swathe of countryside without realising that it is burrowed out with these ancient highways.

The obvious answer to why they were made is that they were not; they just happened. Drive your stock down a grassy, sandstone hill and the grass will be trodden away. If this is repeated for a week, the trackway will become a muddy or dusty mess, and within a month you will probably be down to bedrock. Keep this up for 300 years and you will have a Hollow-Way.

Hollow-Ways, then, are the result of human-induced erosion. Cattle and sheep, taken perhaps to drink in the river or to graze on the water meadows, and people going to and from the fields or passing through, all had a hand – or should I say foot? – in their creation. However, it is the cartwheel that has had the greatest effect. With shod horses and a heavy-laden cart, they would slowly grind away at the land, dislodging a stone here and there

and turning sandstone and chalk to dust. Rain comes to soften the ground and, ultimately, wash the loosened material away. I see this washing away down the Drift, an old drove road into my Dorset village. After heavy rain the flat, tarmacked area at the bottom is covered in chalky mud and scattered with a thousand dislodged flints. Where the geography and geology are suitable, a Hollow-Way will transform itself into a part-time stream and occasional deluge. Should it begin to take too enthusiastically to the idea of becoming a stream, a ditch is sometimes dug alongside it to keep it passable.

The typical Hollow-Way is a sunken path (indeed, this is another name for them) that has simply happened by being used. If it is excavated on purpose, then it is not a true sunken path. Many roads that look like Hollow-Ways are not, because they are made that way by engineers, but any that were Hollow-Ways before being surfaced for traffic are still considered to retain their status. Many paths, Hollow-Ways or not, will have a bank on one or, more probably, both sides. This is typically the case when the land on either side is in the hands of different owners or is in different parishes or manors. Some of the bank material may have been dug out to form a hollow pathway, but the path is likely to become a 'true' Hollow-Way from the traffic it will carry. They are often hedged and can even traverse a wood. In general, if a path seems to have been excavated when there was no need to put in so much effort, then it will be a Hollow-Way.

Hollow-Ways in areas with little hard, rocky material, such as in sandstones, are sometimes strengthened with beach pebbles, other small rocks and, as I have seen several times, builders' rubble. This stops much of the wear and tear, but water will still carve out a channel if it needs one. Those that have been surfaced, usually with tarmac, will obviously not get any deeper. All unsurfaced Hollow-Ways are a species of 'green lane', though, with their shaded aspect and lack of grass verges, they are not always as green as one might expect.

Hollow-Ways have been around for millennia, wherever man has repeatedly crossed a soft and usually hilly landscape. By their nature all are old, and most still in existence date back many hundreds of years. Their very name is commemorated in the surname 'Holloway', which means that a family was named from its home being near one or after the name of their settlement – there are several Holloways in Britain. Either way, the name indicates an Anglo-Saxon vintage at least. But it is worth noting that the surname is sometimes derived from 'holy-way', a path of pilgrimage. Even though the Anglo-Saxons used *hunel-hege* for any tunnel-like woodland path, it is likely that the more specific *hol weg*, 'a road with a hollow', is more appropriate.

Hollow-Ways are, above all, exquisite places to take a walk: hidden, green with dappled light and silent apart from the call of the birds above you, they are reminiscent of the jungle. The best ones I know are not too far from my home. They are the beautiful but decidedly creepy Hell Lane and Shute's Lane ('shute' is yet another word for 'Hollow-Way'). Cutting through the sandstone between Symondsbury and Chideock in West Dorset, they are deep, cavernous places. The walls reach a towering 35 feet high, and the sloping, sandy soil below the vertical sides is covered with plants typical of woodlands such as Hart's Tongue and Male Ferns, Bluebell, Lords and Ladies, Dog's Mercury, Redcurrant, Ivy and the inevitable mosses. The overhanging trees are chiefly ash, oak and sycamore, the ashes sometimes bearing the huge bracket fungus, the Dryad's Saddle. These trees live a precarious life, growing as they do at the top of a crumbling precipice, their roots dangling, and with the occasional tree sat on a shelf-like projection of a tough, rounded, sandstone concretion known as a 'dogger'. Trees here that have given up the unequal struggle can be seen fallen across the Hollow-Way like a bridge. Doggers too can fall and are occasionally found on this narrow path, reminding the walker that it is not just trees that are in peril.

In some places the sandstone that forms the walls has been carved by visitors. Most are inscriptions of the 'Sandra loves Shane' variety, and they are frequently provided with a date. Sadly, while the love between Sandra and Shane may be eternal, the ephemeral nature of the stone ensures that the life of the inscribed declaration will be brief. The oldest one I could find on a visit in 2020 was dated nineteen-ninety-something, though I know that some are older. More artistic productions are also to be seen. Some of these are rather frightening depictions of death's-heads and gurning witches; there is a fading owl, the man in the moon and an ironic carving of a camera. There is a recent abstract relief, about 15 inches high, clearly made by a trained craftsman, which would happily grace a church.

These two lanes are moderately famous for being the model for the fictitious hiding place of the nameless hero in Geoffrey Household's superb 1939 thriller, *Rogue Male*. Much effort has been expended over the years to discover the precise location Household had in mind, but, as in the novel, all his trails are false.

Drama, fictitious or real, is bound to find a home in so romantic an environment as a Hollow-Way. Numerous reports over the last 300 years or so reveal how useful they can be in war. If you wish to hide a company of soldiers in (apparently) open countryside, you can do no better than to pop them into a Hollow-Way. An entire battalion could be moved for a mile or two without any chance of being spotted. Relatively shallow Hollow-Ways form natural trenches and are easier than the deeper ones to defend if needed or to launch an attack from if that seems like a good idea. Unfortunately, while they no doubt work well most of the time, war is a tricky business, and in reports from Waterloo, Sebastopol, Ebelsberg and elsewhere, occupied Hollow-Ways often end up occupied only by corpses. Henry I took his army through a *hunel-hege* after the capture of Bridgnorth, but it was harried by archers hiding in the trees above. Henry later had the trees and much of the surrounding woodland felled and burned.

Smaller-scale disasters have also attended Hollow-Ways. In 1814 a man who was walking across Saltisford Common in Warwickshire heard a cry of 'Murder! Murder!' and found the supposed victim lying in a nearby Hollow-Way. The 'victim' grabbed his Good Samaritan, shouting, 'Now I have you!' The Samaritan gave his attacker a bloody nose and made his escape with nothing more than a hole in his hat from a badly aimed pistol shot and a determination never to help anyone ever again. In 1790 two highway robbers went for the time-honoured method of a stretching a rope across a Hollow-Way. The unfortunate W. Bolton found the rope gripping him in the belly and was quickly unhorsed when one robber whipped the creature back into motion. Bolton was robbed at gunpoint of about £10. Less worrying was the Wombwell Ghost. Its favoured manifestations came in the form of loud screeching which, in 1929, terrified some passing miners in the Hollow-Way near Wombwell Main colliery in Yorkshire. The ghost was clearly being more careful this time as its two-years-previous incarnation as a white-shrouded figure crying 'I am the Monarch of the Wood!' was terminated when it was hit on the head with a pit bottle.

Hollow-Ways are famously adept at filling up with snow and becoming dangerously impassable during heavy rains. Several people have met their end falling into the inescapable snowdrifts they sometimes conceal, and a few have been washed away in floods. Such perils still exist, but the more prosaic dangers of today include the small proportion of careless or incompetent cyclists. Signs asking them to beware of pedestrians along the straighter part of Shute's Lane is a sure indication of past accidents. Off-road motorbikes and four-wheel-drive vehicles can also be an issue. I recently had to squeeze myself against the side of my local Hollow-Way as a car driven by a barely visible ten-year-old boy (accompanied by his proud father) went past. But this is Dorset, so I just smiled encouragingly and waved.

Lime Kilns

He burned the bones of the king of Edom into lime.
Amos 2:1

Lime Kilns are incongruous structures, appearing in unexpected places: lost, vegetation-draped temples of South America hidden in the British countryside, their frequent habit of being half-buried in a hillside completing the picture. Like many old and abandoned industrial constructions, they can be found in suburbia as much as on a remote hill or a canal-side. Lime Kilns are usually noticeable as tall, rectilinear or circular brick buildings, lacking any type of window and possessing what looks like a fireplace where one would expect a door. This is the view from 'ground level'; if you view one from the top, they appear as large brick-lined holes at the edge of a small cliff. They were once in very common use but have now largely been replaced by vast installations producing vast quantities of lime.

If you heat limestone to anything above 825°C you will end up with quicklime. Limestone is calcium carbonate, quicklime is calcium oxide, so there is a great deal of leftover carbon dioxide released during the process, not that anyone worried about this gas in the past. Quicklime is a dangerous material which has always rather terrified me. Having been a keen experimental chemist during my teenage years, I don't normally take fright at lethal chemicals, so I put it down to a sinister television dramatisation of *The Mystery of Edwin Drood*, broadcast in 1960, in which quicklime was used to dispose of a body. It seems that you cannot really dispose of bodies with quicklime, just heat them up and dry them out, but presumably Dickens did not know this.

For the living, however, it is highly caustic, and when water is added it can sometimes produce sufficient heat to ignite anything flammable near by. Breathing in quicklime powder is potentially fatal, drying out and burning the respiratory system. Such an effect has not been allowed to go to waste as it was once employed as an early form of chemical warfare. The technique was to (somehow) launch clouds of quicklime into the air, letting it drift

A Devon Lime Kiln

downwind towards the enemy and praying that the wind didn't change direction.

Fortunately, much more constructive things can be done with quicklime. It can be used in the manufacture of steel and of glass, in the petroleum industry and in many other processes, but it is most often 'slaked'. Slaked lime is quicklime to which water has been (carefully) added, to form calcium hydroxide. This compound is known as hydrated lime or builder's lime. As a building material in lime mortar and whitewash, it has largely been superseded by other materials, though I have a couple of bags of builder's lime in the shed which I use, mixed with selected sands and coal grit (for extra authenticity), to fix the crumbling walls of my old cottage. Lime mortar has been used for millennia and was the only mortar available until Portland cement was invented in Leeds in the nineteenth century (or reinvented – the Romans had something similar). Lime mortar sets hard very slowly, gradually turning the hydroxide back into a carbonate by the absorption of carbon dioxide.

Writing in 1842, the Scottish botanist and architectural innovator John Claudius Loudon stated that 'A Limekiln is a most valuable article on a farm containing limestone, or with limestone in its neighbourhood.' Heathland and clay soils tend to have a very low (acid) pH, leaving them unproductive. Whether and how much of the alkaline quicklime or slaked lime was used historically is uncertain, though lime of some sort certainly was. Quicklime was likely to have been used as it has nearly twice the effect of calcium carbonate (limestone), but it was dangerous, especially as it still needed to be ground or milled into a powder and would certainly burn any grassland to which it was applied. Slaked, it would have been much safer, though its ability to raise the pH was only 25 per cent above that of plain limestone. Both then and now, however, it was plain limestone that was used primarily, though it needed to be milled – a difficult process without heavy machinery, giving the advantage to quicklime, which was soft and very easy to mill.

Lime Kilns have been around for 7,000 years, and a few Roman kilns have been uncovered in Britain, notably a splendid pair of flare kilns (described below) found during the recent building of a bypass to the east of Lincoln. They have been used in Britain over the centuries, though much more so after the Norman Conquest when wooden buildings were replaced by those of brick and stone. The kilns would be built wherever they were required, often on the site of the building for which the lime was needed. Transporting anything very heavy and very far was extremely difficult before the canals and railways were built, so, ideally, kilns were built where limestone was readily available. The lime used to build my cottage is very likely to have been sourced just up the hill from our village, where there is both lime quarry and kiln. In many cases the kilns were used once, then dismantled or forgotten, with only a circle of scorched earth and perhaps a circular hole in the ground to mark their passing.

These early, single-use kilns were similar to charcoal clamps, where wood was partially and very slowly burned over a few days. In Lime Kilns, wood, bracken, gorse and anything else flammable that came to hand was stacked in alternate layers with fist-sized lumps of limestone, then covered with turf and then clay. It was essential that it was ventilated throughout, most particularly through a small, hearth-like opening at the bottom and a hole at the top, so that every part of the mass inside the kiln was assured of complete and even burning. After about three (very smoky) days, the lime was ready, and another couple of days were needed for everything to cool down. The powdery lumps of quicklime would then be sieved out of the potash (the ashy substance left from the wood), or, if it was to be used as a soil dressing, the whole lot would be taken out, potash, quicklime lumps and all. A more primitive system still was to make a bonfire with lumps of lime scattered among the wood.

A more systematic and less wasteful method than the clamp was the 'flare kiln'. From above they look like two-and-a-bit-metre-diameter keyholes, the projecting part being a level or downward-sloping access ramp to the hearth where the fire was lit and the quicklime unloaded when ready. Internally, at least, they were funnel- or half-egg-shaped at the bottom and cylindrical, or with the other half of the egg, above. Towards the bottom of the egg there was a circular ledge on which a rough, temporary vault of stone was built to support the limestone that was to be fired. The limestone was loaded from the top. Beneath this vault was the hearth, which was kept stoked with wood (sometimes coal) for the duration of the firing. The resulting separation of lime from the ash provided very clean lime. Externally, they could be almost any shape or partially underground. They were made from brick or stone and lined with a layer of brick.

Flare kilns have a very long heritage, the Roman kilns mentioned above being of this type, and they are still used today in some parts of the world. They are, however, still *batch* producers, so a continuous method was eventually developed to save the time, fuel and uncertainty involved in repeated, individual firings. These are known as 'perpetual', 'running' or 'draw kilns' and are the kilns described at the beginning of the chapter – and the type most likely to make an appearance on a country walk, as they are robust structures and relatively recent. These kilns were tall constructions which held layer on layer of coal and limestone, supported at the bottom by an iron grid or an arch of bricks. As it burned down, more layers were added. The disadvantage, of course, was that the lime would be mixed with the ash, leaving it nowhere near as clean as that produced in a flare kiln.

With both flare and draw kilns, the fuel and limestone were added from the top. One common characteristic of these kilns is that, if they are situated in an uneven terrain, the shape of the land is utilised to facilitate access to the top. Thus, many are built backing onto a small cliff or slope with the hearth at the bottom. In flat areas, a ramp would be built, as with the kiln on the flat top of Portland in Dorset.

Lime Kilns, like the quicklime they produce, have always been dangerous enterprises, and old newspapers abound with grim stories of vagrants sleeping near them overnight for warmth and being overcome by the inevitable extra by-product of this process, carbon monoxide. Sometimes they would fall, unconscious, into the kiln and be found incinerated on one side. The operators of these kilns could also succumb to carbon dioxide, which is toxic in relatively high concentrations. Worse still, in 1818 an unfortunate farmer from Crosby Ravensworth in Cumbria made the fatal mistake of investigating why the new cartload of limestone had not fallen into the kiln as it should. He gave it a poke and was engulfed in the ensuing avalanche, and trapped in the hot furnace.

Cuckoo Spit

Life is mostly froth and bubble
Adam Lindsay Gordon, 'Finis Exoptatus'

Most people have had the experience of a sudden coldness on their hand or leg during a spring walk or wander around the garden. The mind is sensitive to such surprises, and we may start as though a horsefly has just landed on our skin. Examination of the affected area will find a wetness, and we realise that it was just

Cuckoo Spit. A quick look around will likely find some more, still attached to a plant, gently bubbling away.

Those who have taken an interest in such things might know that Cuckoo Spit contains an insect or, at least, a stage in the life of an insect, known as a froghopper (also called a 'spittlebug'). If you gently remove some of the spit, you will see the tiny (usually green) nymph looking back at you reproachfully and askance (they look at everything askance because their eyes are positioned like the headlights of an E-type Jaguar).

These little creatures go through five instar stages, all tapping into the xylem of the plant and consuming the sugary but low-nutrient broth that emerges. To improve their diet, they engage the services of gut bacteria to make the necessary nutrients. As well as feeding the nymph, this sap is, of course, employed to blow bubbles. Producing bubbles does not seem like a difficult requirement, but nothing, absolutely nothing, in biology is ever simple.

The high-pressure sap forces itself through the nymph. Whatever is left by the time it reaches the lower gut is joined by a cocktail of chemicals supplied by the Malpighian tubules, which are connected at that point. Some of these chemicals partly determine the structure of the bubbles, while others make it bitter as an additional deterrent to predators. Two of the structural compounds are stearic and palmitic acids, more familiar as soap! The completed mixture exits from the rear end of the nymph (which is stuck up in the air) and subsequently runs down the underside of the abdomen. Here there is a cavity into which air is pumped by a bellows-like motion of the abdominal plates either side. The resultant bubbles are wriggled into place to cover the nymph completely.

The froth is, of course, for protection or, as the *Lady's Own Paper* of 1850 more eloquently described it, 'to cover and protect its wingless infancy'. But it is also for hydration. The ancestor of these creatures lived in the soil, tapping the roots of various plants and more at home in that damp environment. The froghopper reproduces this environment high up on a stem.

There are only nine species of froghopper that produce 'spit' in Britain, all members of the Aphrophoridae (zoological families, such as this, are always indicated by the ending -idae). The only other British froghopper is the root-feeding *Cercopis vulnerata*, which makes up for its lack of interesting froth by being black with red markings worthy of a Hell's Angel. The (frothy) Common Froghopper, *Philaenus spumarius*, is unsurprisingly the one you are most likely to encounter. It owes its relative success to not being a fussy eater, establishing itself on numerous plant species, many of which are found in gardens. *Neophilaenus lineatus* is another common species and found on grasses. Froghoppers are related to cicadas and look a lot like them, but are much quieter.

With its flattened body shape and decidedly lateral eyes, the mature insect earned its name from its vague similarity to a frog. It also shares one other froggy characteristic: athletic prowess. Acceleration at take-off is 400 g and the bug can reach a height of 70 centimetres, a remarkable feat that makes it a world leader in its height-to-size ratio, beating the flea by an impressive factor of sixty.

The name 'Cuckoo Spit' needs some explanation. The sixth-century Spanish scholar and cleric Isidorus thought that cuckoos created grasshoppers from their spit, but it is most likely down to the spit being first seen when the cuckoo is first heard, in spring. Nevertheless, some have taken this whimsy seriously, and I found a reference to a mid-nineteenth-century gardener who hated the things because they wilted his flowers – which they do. He spent many of his spring days in the garden, ready to shoo away any errant cuckoos. No cuckoos appeared, but the Cuckoo Spit did.

Galls

Nutkin gathered oak-apples – yellow and scarlet – and sat upon a beech-stump playing marbles, and watching the door of old Mr. Brown.

Beatrix Potter, *The Tale of Squirrel Nutkin*

What do mycologists and botanists do when they retire? They take up cecidology, the study of Galls. Or so it seems. Having enjoyed a lifetime with fungi or plants and learned most of what they want to know, they look for a new interest to entertain them in their twilight years, and it is often the study of Galls that takes their fancy. Several of my mycologist friends have followed this path and bring a wealth of experience to bear on their new interest. With botanists it is an obvious choice because

A mature Oak Apple, caused by *Biorhiza pallida*

they will have seen thousands of them on the plants they study. If there is one constant about Galls it is that they occur on plants, and are, indeed, *part* of a plant. Certainly, there is enough to keep everyone amused, with over 3,000 Galls found in the British Isles, though this is more properly expressed as over 3,000 species known to cause Galls in the British Isles. I seem to be going the same way as my friends, becoming excited every time I see one, even if it is a type I have encountered a hundred times before. Anyway, it is better than wood-turning (see p. 157–8).

But what *is* a Gall? Settling on a definition has proved difficult, as there always seem to be exceptions that clearly *are* Galls but don't quite fit the definition, or exceptions that *aren't* Galls but do! The definition with which few, I trust, would disagree is as follows:

> *Galls are structures on or modification to plants, formed for the purpose of nurturing the offspring of another organism and built or formed by the plant on which it grows under the chemically supplied instructions of said organism.*

Not a very elegant definition, but I think it will do for now. Leaving such essential cecidological technicalities aside for the experts, Galls are the odd protuberances, swellings, leaf curls, fuzzy bits and lumps that appear on various parts of a plant that are caused by the supplier of those instructions. Visually, they can be anything from a 3-millimetre-long bulge on the underside of a leaf to a brown papery mass the size of an apple, to a Sputnik-like conglomeration of wood and twigs a couple of metres across. The covert organism that supplies the instructions and for whose benefit a Gall is constructed (known in the trade by the unappealing name of 'galler') could be almost anything. Insects (mostly midges and diminutive wasp species) are by far the commonest Gall-inducers, followed by mites, fungi, nematodes, bacteria, viruses and even other plants. Although some Galls have common names, one that is applied to the Gall itself, all the Latin names used in referencing them are those of their galler.

If there is any grand differentiation that can be made between Galls, it is in their degree of complexity. This runs a continuum from the undifferentiated mass that makes up bacterial Galls such as Crown Gall to the highly structured and symmetrical Galls induced by Gall wasps. Nearly all gallers are parasites, though a few do give something back to their host plants.

It is worth noting that one or two other odd-looking objects in this book are, or are often classed as, Galls. Witch's Broom (see p. 199) is certainly a Gall. Smuts and Rusts (see pp. 168 and 172) are sometimes classed as fungal Galls, though I have serious doubts about this. Burrs (p. 150) are arguably Galls. All these are sufficiently conspicuous, sufficiently different and often just sufficiently *large* to deserve their own treatment, while those described here are a little more discreet by nature. Most of what follows concerns the commonest of Galls, insect Galls, but first a brief mention of some of the others, as they will be conspicuous to the attentive eye of the countryside walker.

One fungal Gall that I see on occasions is *Taphrina pruni*, the Pocket-Plum Gall. This splendid-looking Gall grows on blackthorn, replacing the fruit (sloe) with a plum-shaped, brown woody Gall, twice the size of any sloe. Moving to the other end of plants, nematode worms are notorious for causing losses to crops as they form damaging root Galls. Cereal crops in particular are affected, with horrific losses throughout history, and still today in some parts of the world. Some nematodes can feed on almost any plant root, with potatoes, carrots, beets, lettuces and even strawberries suffering. No doubt a few gardeners have dug up a failing plant to find dozens of 'knots' along the roots; these will probably be nematode root Galls. Not all Galled roots are 'a bad thing', as the nodes found on some plants, typically members of the pea family, have bacteria-induced Galls that fix nitrogen, which, of course, provides nitrogen for protein synthesis and ultimately fertilises the soil.

Apart from their fascinating shapes and sometimes surprisingly bright colours, complex lifestyle and ecological interest, there is one characteristic of Galls that makes them exceptionally appealing to anyone who cannot rest until they know what everything is called: being host-specific, they are, on the whole, easy to identify. Find a Gall, identify the plant and look up a list of Galls that are associated with it. Match what you have found to one of these. It will also help to note where on the plant the Gall was found – leaf, stem, fruit, root and so on. While it is, obviously, necessary to be able to identify the plant in the first place, this is not difficult, with a hundred books out there to

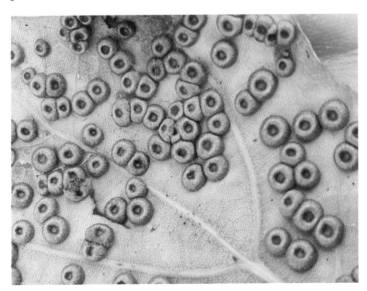

The Silk Button Gall, caused by *Neuroterus numismalis*

help you. It is certainly easier to identify a plant species than the 2-millimetre-long cynipid wasp for which the Gall was made.

Many people will recognise the more obvious and common Galls at least for what they are, though on the fungus forays I lead there is always someone who brings me a Spangle Gall (dozens of green or brown discs on the underside of an oak leaf) or a Knopper Gall (a determinedly lumpy Gall which grows in place of an acorn), asking me if it is a fungus. What few people know is what they are *for*, except that they are something to do with insects. Fewer still know how they are created, and in this I am including professional researchers, who have only just begun to unveil their secrets.

Among the prettiest of all the Galls is the brilliantly red and spiky Robin's Pincushion Gall, made for the Gall wasp, *Diplolepis rosae*, that induces it. It grows on the Dog Rose, *Rosa canina*, and a few other *Rosa* species. In early summer, the adult wasp lays about thirty eggs on a terminal or axial leaf bud, burying them slightly in the plant tissue using her ovipositor. She will visit several sites in her short lifetime, laying an average of 500 eggs. The eggs, which are deposited close together on any individual site, each initiate the construction of a Gall by stimulating plant tissue growth using chemicals that pass from the buried end of the egg through the permeable shell. Neighbouring, growing Galls merge into a single, multi-chambered Gall, each containing a single larva. The larva develops through those many stages beloved of insects to emerge as an adult the following late May or early June. This is a fairly typical galling process, but it can be more complex.

The most familiar of all the Galls is the Marble Gall, found attached to the twigs of oak trees. It is brown, woody when mature, and usually with a little hole drilled in it. Cut in half, it will reveal a tough skin surrounding a corky mass and either a hole in the middle – or the two halves of an unfortunate cynipid wasp larva whose blood is on your hands. The Marble Gall is so called because, well, it looks like a marble and is the size of a marble. The inducer of the Marble Gall is *Andricus kollari*, the Oak Marble Gall Wasp.

The complexity with this Gall is that if you look it up in your handy book of British Galls, you will be directed to two separate pages for the same species of gall-inducer (*A. kollari*). One will

Robin's Pincushion, caused by *Diplolepis rosae*

describe the Marble Gall on Britain's native oaks, *Quercus robur* and *Q. petraea*, and the other a (different) Gall on the introduced Turkey Oak, *Q. cerris*. The latter Gall is tiny (2 millimetres long) and bullet-shaped, found again on the tree's twigs rather than on larger branches. *A. kollari* has alternate sexual and asexual generations, the Marble Gall being the latter. This Gall is so well known that it seems to be an essential and ancient feature of the British countryside, but the Turkey Oak only arrived here in 1775 and the wasp, travelling in a consignment of Galls destined for the dyeing and ink industries, in 1830.

Although there are tens of thousands of Galls worldwide, there are many, many species of insect, fungus etc. that do not form them. What, then, are the advantages of building a Gall? There are two considerations: protection and food. The two Galls just discussed are clearly able to keep their tasty grubs safe from all but the most determined of predator, and all Galls provide food. A Gall is a fortified dining room. The larvae effectively shape their chamber by wriggling about and eating parts of it. The food supplied, known as 'nutritive tissue', provides a perfectly balanced diet, consisting of high sugar levels, lipids, protein and more, all contained within plant cells that are 'hypertrophic' (overgrown). Considering that this is an entirely closed space, one must assume that it will include 'facilities', but a scrupulous search of the literature has yet to discover any academic paper on so obscure a subject.

An impenetrable fortress it may seem to be, but Galls are nevertheless surprisingly cosmopolitan places. This is mostly down to 'early adopters' – that is, species that find their way into the *immature* Gall and bide their time. Some are little more than a nuisance, taking up space and food – the 'inquiline' species, which move uninvited into someone else's home. These are often various wasp species (again, very tiny) or aphids that are attracted to the volatile chemicals produced by the host plant or the Gall. A few small beetles, such as *Mordellistena convicta*, will take up residence, but by adopting the more direct approach of boring into the mature Gall. More troublesome invaders are the parasitoid wasps: members of the Ichneumonidae. 'Parasitoid' creatures always kill their host, whereas simple parasites just take advantage. It is not only the species that built the Gall that suffers from these wasps; most inquilines can succumb to them too. Then there are 'hyperparasitoids', which feed on parasitoids. A Gall can thus be a very busy place and, with all these interlopers, only a few Galls will produce a viable adult of the expected species.

How Galls, the complex ones at least, are created remains an unsolved puzzle. It is known that a cocktail of chemicals, usually plant hormones, used to stimulate rapid cell growth and division is produced by the galler. The ability to produce some of these chemicals appears to have been borrowed from symbiotic bacteria within insects via horizontal gene transfer. How it is controlled to produce novel plant tissues and complex structures with everything in the right place is still open to speculation. A particularly exciting idea is that the insect uses bacteria to change the behaviour of the host plant. It is known that the Crown Gall bacterium, *Agrobacterium tumefaciens*, transfers plasmids (bits of wandering genetic material) into plant cells to effect and control the growth of the Gall. It is hypothesised, then, that those gallers that produce the more complex Galls use horizontal gene transfer via a bacterial intermediary to transfer genetic instruction sets into the genetic machinery of their host. It may be simpler than this seems, for many Galls look a little like fruit or flowers, so a redirection of genetic expression in cells already primed to create these structures may be most of what is needed.

Dodder

The red weed grew tumultuously in their roofless rooms.
H. G. Wells, *The War of the Worlds*

Finding an area of gorse covered with a mat of leafless, bright red threads is quite startling. The mat seems to hover over its host plant, covering it, choking it of life and always reminding me of the Martian 'red weed' from *The War of the Worlds*. Dodder, *Cuscuta epithymum*, is a parasitic plant which leads a life of shame and dependency by sponging off its fellow plants. It contains little or no chlorophyll, instead sucking nutrients from its host. It is one of three British species, the other two being Yellow Dodder, *C. campestris*, and Great Dodder, *C. europaea*. These are less common, though none of the three is encountered frequently.

Dodder in flower and looking its best

They have a generally southern, lowland distribution in Britain, though they are also found in Yorkshire and central Scotland.

Despite their relative rarity, some species are frequently classed as 'invasive'. Indeed, they can become a serious nuisance where they have established themselves as they are the devil to disestablish. Although they are all annual plants, they produce seeds with thick coats which form an intractable seed bank, with seeds germinating at random intervals over the years, like exploding unexploded bombs.

Like most parasites, they are choosy about which plants they will parasitise. In the case of the Dodders, the choices are limited but surprisingly eclectic. Dodder prefers heathers and gorse; for Yellow Dodder it is carrots that are top of the menu; and for Great Dodder it is stinging nettles. This is already a wild ride through the plant families, but many other plants can fall victim: beets, raspberries, mints, willows, hops and, oddest of all, onions.

The seed germinates (eventually) and sends up a short (up to 8 centimetres) shoot, using only the energy within its seed. It does not do this randomly but is attracted to the far-red radiation transmitted by leaves that are rich in chlorophyll, thus choosing more mature plants.

When contact is made, it wraps itself around a stem and quickly forms what are known as haustoria. Externally, these look like caterpillar legs growing from the Dodder stem. The haustoria extend hyphae into the host and plumb themselves into the vascular system. Once mature, the Dodder plant produces huge numbers of small, clustered, pale pink, yellow or orange flowers, each one producing a single seed.

The Dodders' habit of entwining themselves around other plants gives a clue to their heritage. Although they are all placed in a subfamily, the Cuscutoideae, this is within the family Convolvulaceae – the bindweeds. Indeed, all bindweeds, including the Dodders, spiral like a corkscrew, but some plants spiral the other way. There is a song by Flanders and Swann entitled 'Misalliance', about the romance between a right-handed bindweed and a left-handed honeysuckle. It ends badly.

Broomrapes

[...] where in the Castle ruins grew the Ivy Broomrape, and on the walls of the old city flowered the wild Carnation.
    Sunday Magazine, 1880

We all have expectations of the natural world. Snakes should slither, wildebeest should be chased by lionesses, wasps should be scary and leaves should be green (or, at the very least, red). What you do not expect is plants that are entirely pale pink or living-room beige.

Forty years ago, in the hazel coppice near my home, I found one of the oddest plants I have ever seen. There were dozens of them, all clustered around the base of the hazel stands. Their most noticeable feature was that they were entirely a dull pink, with not a hint of green in sight, and they lacked leaves, consisting only of a scaly, curling stem covered in flowers. Overall, they had the appearance of large, upright, cooked prawns. They were Toothworts, *Lathraea squamaria*, one of the members of the Broomrape family, the Orobanchaceae.

Only about a dozen of the Orobanchaceae are found in Britain, two being Toothworts (*Lathraea*) and the rest Broomrapes (*Orobanche*). Of these, only two could be described as common: Toothwort and Common Broomrape, *Orobanche minor*. The very bright and beautiful Purple Toothwort, *L. clandestina*, is found only occasionally, and Ivy Broomrape, *O. hedera*, is locally frequent but mostly on southern and western coasts. The remainder are very rare and generally considered to be non-native.

If you have never seen a member of the Broomrape family, I strongly advise a couple of days wandering around the Isle of Wight. Either there is an Orobanchaceae enthusiast living there,

Carrot Broomrape, *Orobanche minor* subsp. *maritima*

filling every waking hour recording every specimen he or she finds, or it is Broomrape Central. I have only ever seen Toothwort and Common Broomrape, the latter parasitising Wild Carrot just above a Dorset beach, but I live in hope of finding more. They are very distinctive with their deathly colours and are quite tall, so they should not be hard to find if they are around.

Since you are only likely to find the four above-mentioned species, identification should not be a problem, but a quick

Toothwort

survey of (very) nearby plants will help to some degree if you are stuck: Toothwort with hazel, Purple Toothwort with willow, alder and poplar, Common Broomrape predominantly with clover and other pea family members and, rather obviously, Ivy Broomrape with ivy. If you want to make life a little more difficult for yourself, then, if you find a Common Broomrape, try to determine which variety it is. My Dorsetshire Common Broomrape, in growing by the sea and on Wild Carrot, was likely to be *Orobanche minor* var. *maritima*.

Unlike the distantly related dodder (see p. 146), which attaches itself to the stems of plants, Broomrapes attach themselves to the roots. The Broomrape seeds produce roots only when first searching for a host. They will not attempt to connect to a plant that is not a potential host as they can identify the correct plant by its chemical signature. These chemicals are known as 'xenognosins', where 'xenognosis' means, loosely, 'knowing the foreigner'. Once the root detects these, they form haustoria and penetrate the roots of the host plant. Interrupting this process has been the aim of researchers hoping to reduce the quite serious agricultural losses incurred when Broomrapes attack crops.

Burrs

THE CHOICE VENEERS in fine condition are of rosewood, black ebony, tulipwood, English, American and Hungarian ash, walnut burrs, elm, Dantale oak, brown oak, birch, Spanish mahogany, &c.
Advertisement in the *Liverpool Mercury*, March 1887

Anyone who wanders the deciduous woodlands of Britain or has ventured into a venerable park will have seen a Burr. They are the lumps that grow on the sides of a tree and are known as 'burls' in

North America. Most often it is the trunk that bears them, and frequently near the bottom. All Burrs are covered with bark, though it will be very coarse and lack linear fissures. The tree itself will probably be healthy, its Burr worn with fortitude and as a proud mark of maturity.

For about thirty years I worked as a reasonably accomplished country cabinetmaker. I would buy local timber, some purportedly sustainable timber from abroad and veneer. Veneer has acquired a bad name due to its use in cheap, particle-board

Large Burr on sweet chestnut

furniture, but it has a long history in the construction of high-class furniture. The most expensive of these veneers are the 'Burrs'. I would drive to my veneer merchant in Curtain Road, Shoreditch, and spend a happy hour or two clambering around the tumbledown building, which looked like one of the sets for the film *Oliver*, though Fagin never did appear. There were some very expensive exotic Burrs, but the most costly European Burr was usually walnut. These days it is around £100 per square metre. It would always be sold in multiples of four consecutive leaves, enabling the cabinetmaker to 'quarter' them to produce a symmetrical pattern both top to bottom and side to side (think of two butterflies, wings open and placed bottom to bottom).

In addition to being sliced for veneer, Burrs are used 'in the solid'. The best of these generally fall into the reliable hands of wood-turners, and some of the worst into the seriously unreliable hands of people who make, and presumably like, chunky, rustic tables with bits of branch as legs. Good-quality Burrs are extremely valuable, and there are many recorded thefts of them from important trees, such as the giant redwoods in the US. To be honest, whenever I see an oak Burr, I feel the temptation to wield my chainsaw, but wiser counsel has always prevailed. Trees damaged so drastically seldom recover.

Nearly forty years ago someone called me to ask if I would like a burred elm that had blown down in his garden. I went to see it and found it to be covered in dozens of Burrs, some of them enormous. After expressing much gratitude to my benefactor and a bit of necessary chainsaw work, I loaded it into my pickup and took it to the local sawmill. Over the years I have made many things with this glorious elm tree. Most have been small, but I did use some for panels in the construction of a kitchen, the frames around them made from slightly burred but mostly straight-grained elm. It may seem something of a waste to use so precious a timber in a mere kitchen, but it was to go under a massive elm plank which gets a mention in Domesday! Much to my wife's distress, I still have some left under a tarpaulin in the garden, and my favourite piece (which is far too good to make into *anything*) is pictured on the next page.

Burrs are usually referred to as 'tumours' because that is precisely what they are – excessive growth of unnecessary material. Fortunately, they do little harm and are therefore regarded as benign tumours. Sometimes a trunk will be covered in scores of small Burrs; sometimes they are huge at a metre wide and high. The latter are highly prized by woodworkers and

Large sphaeroblast with convoluted wood visible

Elm Burr, cut to show the intricate patterning

veneer manufacturers; the former just produce planks of wood that are 'interesting'. As my one-time Devon timber merchant would say, 'I got some nice oak here, John. Tis a bit pippy, mind,' meaning that small Burrs were scattered over the otherwise straight-grained timber like paw prints. Sometimes, Colin's 'pippy' oak was also a beautiful ruby brown rather than the usual straw colour, meaning that it had been infected by the Beefsteak Fungus, *Fistulina hepatica*.

Burrs are produced by trees under stress. Various stress factors can be blamed (or thanked) – viruses, fungi, insects, fire and mechanical damage – though discovering which one of these applies in any particular instance is not usually possible. The stress stimulates hormone production, which causes rapid and uncontrolled growth (hyperplasia) of xylem tissue (the plumbing system which transports water and minerals from the roots). If you look closely at a slice of Burr, it seems as though thousands of shoots have rushed for the exit, growing densely packed together to form a solid mass of tiny knots. In fact, they are dormant buds.

Technically, Burrs are a sort of 'sphaeroblast', meaning 'throwing young growth'. However, the word is more usually applied to lumps, both large and small, on trees that are entirely smooth. Inside these growths, the wood is not made up of lots of shoots but of convoluted normal growth. These are common on beech trees, and the beech sphaeroblast pictured on the previous spread helpfully displays the convolutions.

There are a few other lookalikes that can be confused with Burrs. Cankers are similar but generally smaller and less well structured, sometimes looking as though something has exploded outwards through the bark. They are caused by fungi and, sometimes, bacteria, and there may be dozens of them on a single trunk and along the larger branches. Witch's Broom (p. 199) and recently cut pollards (p. 117) are two more vague lookalikes. Crown Galls are, however, the most likely to cause confusion. These are infections by the soil bacterium *Agrobacterium tumefaciens*, which cleverly transfers bits of its DNA (plasmids) into the plant tissue to cause abnormal growth. This growth tends to completely surround a stem, disrupting vascular systems. They seldom grow very large, no doubt because the plants are usually dead before they can do so. These galls are strangely sinister in appearance and effect. Crown Galls are found on many plants, including hawthorn, willow, roses, fruit trees and, unexpectedly, beetroot.

Spalted Wood

As some black fortress rolls the headlong tide of war's impetuous flood.
'An Englishman', *English Work and Song amid the Forests of the South*, 1882

When I was a furniture-maker, people (well, men) would come to my workshop, admire my work and then proceed to tell me that their retirement plan from their job as a hedge-fund manager was to take up wood-turning. I was always very encouraging,

Spalted beech logs

and indeed it is a wonderful skill to learn and a pleasure to do, but my heart always sank for their friends and family. Turned dishes, candlestick holders, little boxes, random ornaments and a thousand other turned objects are terrific fun to make but a

blot on the domestic landscape when on display. With a few honourable exceptions, they look tacky. I have turned thousands of knobs, chair-stretchers and table legs, but these were functional and not destined for some poor devil's mantelpiece.

Despite still being of the opinion that most turned objects should have been strangled at birth, I have been hypocritically eyeing a bigger, better lathe and, most fun of all, collecting bits of wood that would look nice as a vase on Cousin Fay's mantelpiece. Prominent among these are some beech logs rescued from the wood pile that feeds our stoves. Most beech logs are plain and dull, but a fair proportion have black lines running through them which are visible wherever the log is split. The wood is 'spalted'. In fact, and rather obviously, these lines are really internal surfaces, as can be seen or at least inferred from the photograph of Fay's vase. I know she'll love it.

While a pile of domestic beech logs is by far the most likely place to clearly see spalting, the black lines are sometimes visible on formerly internal surfaces, such as on broken tree branches and within split forest trunks. Also noticeable may be colour and textural changes between the delineated zones: deep cream to pale cream, occasional bright greens or red, and hard wood to soft and spongy wood.

When a tree dies, the moment its moisture content is sufficiently low, any fungi that have already set up camp inside the wood waiting for their moment spring into action. Beech Tarcrust, *Biscogniauxia nummularia*, which forms black patches on beech bark, Candlesnuff, *Xylaria hypoxylon*, Cramp Balls, *Daldinia concentrica*, and Brittle Cinder, *Kretzschmaria deusta*, are four such fungi. Their mycelium grows and feeds and keeps going until it runs out of enthusiasm or suitable wood, or until it meets another mycelium. If the latter, it may be a different species or a different instance of the same species. Some fungi are combative, destroying all other fungi in their wake, but many come to an amicable arrangement with their neighbours. Like all good neighbours, they build big fences. Sometimes a single fungal colony will simply stop growing and put up a fence anyway, presumably thinking that you can never be too careful.

The black 'membranes' are known as 'pseudosclerotial planes' (PSPs), and they will entirely, if irregularly, encompass the mycelial colony. When two mycelial colonies meet, they produce these PSPs, providing an impenetrable barrier between them. PSPs are also excellent barriers to water, ensuring a reasonably steady environment for the mycelium. PSPs are black and made of cells with very thick walls which include melanin.

'Sclerotia', from which PSPs derive their name, are produced by some fungi as survival capsules. The genetic material is kept safely inside a tough covering almost identical in chemical make-up to PSPs, awaiting a time when conditions are favourable. Ergot, *Claviceps purpurea*, is nearly always seen in its sclerotial phase as elongated, black seed-like projections on grasses.

Each colony within its walls will be at a different stage of growth, and perhaps be a different species. This explains the differences between the wood colours and textures. The unclaimed territory will be healthy wood (well, relatively healthy for something that it is dead) – in the case of beech it will be a warm cream. Once a fungus takes hold, it will consume the darker lignin, leaving the wood pale and soft. Too soft – 'punky', as it is known – and the wood-turner will have bits of rotted wood flying all over the workshop. Spalted timber should never be used for structural components such as chair legs; although sound enough for a candlestick or decorative table-top, the wood inside its sclerotial fortress is still rotten. Brightly coloured zones are rare in British timbers, though they exist in North America and are sometimes artificially induced by the introduction of the appropriate fungi into the timber, and even 'ordinary' spalting is sometimes induced.

Beech is not the only wood to form spalting patterns. Birch, maple, walnut and many other woods can display this patterning, though beech spalting is the most common. I have some hefty pieces of boxwood which I should have kept drier than I did, and the markings are very pretty.

Another occasion when you might find a PSP (of sorts) explains the mystery of why hazel branches sometimes stick together. Here it grows *outside* the tree. The pinky-grey, crusty and slightly hairy fruiting body of Glue Crust, *Hymenochaete corrugata*, is a common species of fungus that forms on a hazel branch. When the fungus comes into contact with another hazel branch, it will produce a mass of sclerotial material to stick the two together and then infects the new branch.

Finally, a very common sight on woodland floors is bits of branch, whole branches and, occasionally, an entire fallen tree, which are almost completely black. If you break such a branch

Piece of wood encompassed by a PSP

in half, beneath the black layer you will find the wood to be the typical pale colour of rotten wood, fairly dry and with varying degrees of decomposition and resultant fragility. The black surface does not quite look as though the wood has discoloured through, say, the acquisition of iron compounds or humic acids, rather that it has been given a coat of blackboard paint about half a millimetre thick. These are PSPs which have encompassed and enclosed the wood, protecting the mycelium while it digests the contents. The fungus that performs this magic is usually *Kretzschmaria deusta*.

Squirrel Dreys

> *They made little rafts out of twigs, and they paddled away over the water to Owl Island to gather nuts.*
>
> Beatrix Potter, *The Tale of Squirrel Nutkin*

Squirrel Dreys form one of the several large, rounded objects that can be seen on trees. The others are mistletoe, burrs, bird's nests and larger galls, such as Witch's Broom. Dreys, however, are easy to spot for what they are, as they are usually formed at the crux of branch and trunk and are only perched towards the ends of branches occasionally and by squirrels who are looking for a bit of excitement. They are invariably quite large, at about 30 centimetres in diameter, compact and almost invariably spherical. Crucially, they always include leaves in their construction, something that cannot be said for burrs or galls, though, obviously, mistletoe rather ruins any wider generalisation.

As is very well known, we have two major species of squirrel in Britain, one native and one a late interloper. The latter, the Grey Squirrel, has banished its native cousin, the Red Squirrel, to northern England and to Scotland, and to a few other small outposts. I have seen the Red Squirrel on Brownsea Island off Dorset a few times and it seems happy there, untroubled as it is by its boorish cousin or by dogs, which are excluded from the island. I also saw a Red Squirrel in the middle of Copenhagen a few years ago, running along the fence behind the restaurant I was in. There is also the Black Squirrel, which has established a footing centred on Hertfordshire, and which is believed to be a hybrid species resulting from the unholy union of a Grey and a North American Fox Squirrel. For most of Britain, it is the Grey Squirrel that comes to mind for most people, though all three can produce dreys. However, it is the dreys and the habits of the Grey Squirrel that I will describe.

The Grey Squirrel evokes mixed feelings among the British. It is a usurper of the much-loved Red Squirrel, but very cute. It can damage trees by stripping the bark, but still it is very cute. I could go on, but I am sure you understand my meaning – cuteness goes a very long way. Despite my views on invasive species, I too love the Grey Squirrel. The flavour is very similar to that of the rabbit, if a little nuttier, and I am surprised that it is not considered more of a delicacy. If you ever get the chance, try Grey Squirrel offal kebabs, one of the best dishes I ever ate.

A friend of mine was having trouble with Grey Squirrels nesting in his loft and set traps which killed them instantly (or so he said – at least it would be better than poisoning). Knowing my fondness for squirrel pie, he would bring them round to my house, hanging them in a bag on the front door if I was out of the house. One day when I was out, he found that he had not brought a bag and posted them through the letterbox instead. On my return, I was greeted by a sea of grey corpses on the door mat, and a friend who was with me required a sit-down and a shot of brandy.

Squirrel Dreys would seem simple enough to explain, but they are more complex in nature than generally thought. Grey Squirrels produce not one type of drey but three: summer, winter and breeding. The summer drey is often missed by the walker in the woods because it is a relatively small and untidy pile of twigs and leaves, forming a platform high up in a tree and invariably hidden by leaves. It is a little like the squirrel-raft of Beatrix Potter's imagination except that it is just for sleeping during warm and dry weather. The winter drey is the most substantial of the three and the one that generally gains our attention once the leaves have fallen in November. The breeding (or natal) drey is a halfway house between the other two. Squirrels can also find a home in a tree hollow – a 'den' or 'cavity drey' – but effectively build a drey inside the hollow, using part of the same procedure described below.

Breeding is very dependent on weather and food supply, and can occur once, twice or not at all. In a very good year the first 'kittens' are born in February, and the latest to be born may be in the middle of November. All of the dreys are built in the warmer months, long before leaf-fall, as is witnessed by the persistent leaves that remain attached to the twigs that form the main structure of dreys.

The winter drey is a complex construction, seemingly based on the simple principle of making a huge pile of leaves and twigs (or just twigs), then burrowing into it to make a hollow space and building the more intricate domestic details from the inside. In fact, the squirrels first build a platform of twigs (though they will cheerfully take the short-cut of using an abandoned bird's nest), then arrange the twigs with more care than I suggest, becoming increasingly fussy about twig suitability and positioning as the main bulk nears completion. The layer inside the structural twigs is really multiple layers (up to thirty or so) of leaves, which form interior 'shingles' to keep out the rain. The innermost component is a lining made of soft materials such as dried grass, moss, fur, feathers and wool, all helping to keep the small interior warm. Even in very cold weather a drey can maintain a temperature of 25°C and above. The entrance hole is on one side. Such dreys can last for years, though a certain amount of annual home improvement is required to keep them habitable, and they can even be passed down to subsequent generations. Nest-sharing is also a common practice, though an interloper from a nearby parish is likely to face violent eviction. Squirrel Dreys are true homes, with much coming and going as Grey Squirrels do not hibernate.

Lackey Moth Eggs

'The band! The speckled band!' whispered Holmes.
    Sir Arthur Conan Doyle, 'The Adventure of the Speckled
    Band'

I had decided to avoid any moths in this book, thinking that they formed too big a subject and one lacking anything much that was not immediately and obviously identifiable as belonging to a moth. Then my very mothy friends Christine and Colin sent me the photograph you can see here, pale eggs helically laid around a small branch. I have never seen this remarkable construction before. The Lackey Moth, *Malacosoma neustria*, which produces it, is a common species in southern England and coastal Wales, so I guess I haven't been paying enough attention. This moth is not always welcome in the garden, as roses and fruit trees are among its favoured habitats and food plants. Though not as terrifying as the speckled band that Sherlock Holmes had in mind, the Lackey Moth is ominously classed as a 'defoliating moth': the larvae can

Small Eggar caterpillars

strip a tree bare. With a few fruit trees and several roses in my garden, perhaps I am just lucky in not having seen the eggs before.

The eggs are laid around a small branch of any one of its host trees, usually on the southern side. 'Around' is the important word here as, when complete, it looks like a rather stylish bracelet. The photograph, taken in early January 2020, shows the mass laying of about 250 eggs, and is fairly typical. It also shows a distinctly helical form, with a left-hand thread, though there is no consistency about handedness. Some sort of helix is inevitable if the instruction set followed by the moth is 'lay an egg on a suitable host, lay another next to it at right angles to the orientation of the branch and keep going'. Judging by the remnants of eggs from previous years, it seems that this branch at least was a favoured spot.

The slightly Ridley Scott shape of the individual eggs is typical for moths and butterflies, though few are laid in so artistic an arrangement. The eggs are firmly fixed using a proteinaceous glue which sets rock-hard. Destruction from complete removal now unlikely, the eggs still face challenges, mostly in the dread form of parasitoid wasps. Some are early-adopters, infecting the eggs soon after they are laid in August or September, but most find their way into the eggs over the months before they hatch, some time during late spring and early summer. About one-fifth of the eggs succumb in this way.

The hatched larvae live communally in a tent, suspended from their host. The photograph on the previous page is of the similar (and slightly more impressive) tent spun by the Small Eggar Moth, *Eriogaster lanestris*. Lackey Moth larvae are flamboyant beauties. They have symmetrical, longitudinal stripes which, starting with a single line of white in the middle, run their own rainbow of black, orange, black, orange, black, blue, black, orange, hairy.

Smut Fungi

O Rose thou art sick.
The invisible worm,
That flies in the night
In the howling storm:

Has found out thy bed
Of crimson joy:
And his dark secret love
Does thy life destroy.

William Blake, 'The Sick Rose'

Sometimes there is something 'off' with a plant. It may be nothing more than a random malformation, or leaves strangely reshaped by nibbling larvae, but sometimes it is much more interesting. My first Smut Fungus, years ago, consisted of an area of wheat plants with dark brown, almost black, seeds. Maybe they had just rotted, I thought, and in a way they had, but instead of going squidgy or slimy, they had become dusty. It was *Ustilago tritici*, a Smut Fungus, and the 'seeds' disintegrated into a grubby mass of dark dust on being touched.

Smut Fungi are ubiquitous, though they generally go unseen or unnoticed. They are most abundant on grasses and sedges but also infect several, seemingly random, herbaceous plants, such as potatoes, pinks, teasels, onions and bladderworts, though not all of these are common or even occur in Britain. One of the few that may occasionally be seen in the garden is the Corn Smut, *Ustilago maydis*, though it is quite rare. It is, however, spectacular, forming clusters of spore-filled pustules known as 'sori'. Most of the once-common species have disappeared, notably those that used to afflict cereal crops that have undergone extensive selection to form Smut-resistant strains. The loss of these interesting species is one of the prices we (and they!) have paid for agricultural improvement.

The idea that plants indulge in sexual congress is a fairly recent one. The great eighteenth-century botanist and taxonomist Carl Linnaeus, best known for devising the two-word system for naming organisms (*Homo sapiens* etc.), was not the first to discover that plants were sexual organisms (that honour is generally credited to Rudolf Jakob Camerarius), but he was certainly the most outspoken and notorious exponent of the concept. When Linnaeus published his 'sexual system' for identifying flowering plants by counting their stamens and pistils (male and female parts), there was outrage, though his use of overt sexual metaphors such as 'four husbands (stamens) for one wife (pistil)' was rather asking for trouble. Few believed that so pure and innocent a thing as a flower would or could sully its reputation

Red Campion with infected anthers

in so demeaning a manner. How much more outraged those eighteenth-century critics would have been had they known that flowers can also suffer the 'wages of sin': sexually transmitted diseases.

Enter one of the commonest and most noticeable of British Smuts, the 'anther' smut fungus, *Microbotryum violaceum sensu lato*. Note: *sensu lato* means 'in the broad sense', that is, there are lots of species very closely related to this one, which was once considered to be alone. Although any Smut infection will be present throughout much of the plant, it is usually the reproductive organs (flowers) that are commandeered to produce spores. Among the 'pinks' (members of the Caryophyllaceae family) that are infected by *M. violaceum* is White Campion, *Silene latifolia*. It is a very common plant of north-western Europe and can readily be found in wastelands and field edges, and it is closely related to several other species of Campion, including the very common Red Campion, *Silene dioica*. These too can fall victim to the fungus, and it is a Red Campion that is pictured here. As can be seen, it is the infected anthers (the part of the stamen that produces pollen) that are noticeable, though it takes an observant eye to spot them.

The anthers of White Campion are the typical bright yellow of pollen – most of the time. However, if the plant is infected with the Smut, then the pollen is replaced by dusty black spores (teliospores, to be precise) and the petals become grubby, as though dusted with soot.

White Campion is dioecious: that is, it has separate male and female plants, a way of life adopted by fewer than 10 per cent of plants. 'Dioecious', incidentally, is one of Linnaeus's matrimonial metaphors, meaning 'of two houses'. Sexual transmission occurs through the familiar agency of pollinating insects, which, seemingly unconcerned on finding black 'pollen' rather than yellow, transfer the spores to another, healthy, White Campion. Here the spores will mate with other spores. The fungus will eventually develop hyphae (filaments of fungal tissue), which will grow throughout the plant and feed. Campions are perennial, or at least biennial, so new spores can be formed on the anthers each year. Or so an unreflective reading of this might suggest. In fact, the spores cannot form on the anthers if it is a female plant with only pistils!

Female plants have the potential to be more productive (for the fungus) than male plants, because the flowers on the former will be retained for longer, giving the spores more time to develop and be dispersed, yet female plants produce only pistils, not anthers. How, then, can this possibly work? It is quite simple, but sounds complicated: in uninfected plants the growth of anthers is naturally suppressed in female plants and the growth of pistils is naturally suppressed in male plants, thus providing the correct sexual organs for the genetic sex of the plant. In infected female plants, pistil production is artificially arrested by the fungus, while the suppression of anthers is removed. Effectively, the fungus changes the phenotypical (what it looks like) female into a phenotypical male by allowing it to grow anthers. Males are largely unchanged by this biochemical trickery, though both male and female plants are generally rendered sterile in a process unpleasantly referred to as 'parasitic castration'. Thus not only do plants engage in sex and contract sexually transmitted diseases, but they also practise some form of 'transgenderism'. While the modern world takes a welcome laissez-faire view of the latter, delicate eighteenth-century sensibilities would be turning in their graves.

Rust Fungi

> *Where moth and rust doth corrupt*
>> Matthew 6:19

Unlike my gardening friends, I am always pleased, excited even, to find a plant infected with Rust. Many Rusts are quite startling in appearance, with 'beautiful' not being entirely appropriate for anything that is generally described as a pustule, for pustules they are. Sometimes.

Rusts, like most of the other subjects of this book, go unnoticed by most people, and few will spot them even on the fungus forays I take. But in truth they are everywhere, swelling stems with tumorous orange lumps and turning leaves blotchy, dusty, pustulated and dead. At their prettiest they decorate leaves with brilliant orange 'aecia' (reproductive organs that produce sexual spores), contrasting nicely with the green. The undersides

Rust on Stinging Nettle, *Puccinia urticata*

of Common Mallow leaves are very often covered by scores of these little raised, orange/brown bumps, and on Alexanders there will be large raised patches of brilliant orange. If you have ever wondered why grasses sometimes turn a dirty orange, then this too will be a Rust Fungus.

By far the most obvious Rust Fungus for anyone to see on a stroll through the countryside is *Phragmidium violaceum*. It is this that produces purple blotches on a large proportion of the brambles in the country. If you turn such a leaf over, you may see the bright orange urediniospores; if you see black dots, they are the teliospores. This fungus has been introduced as a biological control agent for brambles in parts of the world where they are neither native nor welcome, though not, it seems, with unqualified success.

Believe it or not, it is quite easy to identify most Rusts, even for the untutored eye. Anyone who despairs of naming even common mushrooms and toadstools will find a relatively easy path before them with the Rusts. Indeed, it is often possible to make a guess with some hope of being correct, though

knowledge of the Latin name of the host plant is required. Over a third of the 300 or so British Rusts are in the huge genus *Puccinia*, which has a worldwide total of 40,000 species. The specific epithet after *Puccinia* is often the generic name of the host plant. So the Latin names of the Common Mallow (*Malva sylvestris*) and Alexanders (*Smyrnium olusatrum*) Rusts are *Puccinia malvacearum* and *P. smyrnii*; that of the splendid Rust that grows on Stinging Nettles (*Urtica dioica*) is *P. urticata*; and if your Antirrhinums are looking a little wan, they are suffering from *P. antirrhini*. The fact that *Phragmidium* species of Rust always grows on members of the Rosaceae is also a help for the budding rust enthusiast.

Rust Fungi parasitise very many plants. As I have suggested, the garden is likely to be where most people encounter them. Here their rose leaves can be discoloured yellow by *Phragmidium tuberculatum*, their Fuchsias rainbow-blotchy from *Pucciniastrum epilobii* and their pear trees ugly and unproductive from *Gymnosporangium sabinae*.

It is interesting to note that, despite their enormous difference in appearance, the Rusts belong to the same division of the fungi as most mushrooms and toadstools, the Basidiomycetes. Fungi in general are notorious for being complex in their feeding and reproductive strategies, but Rust Fungi are simply bizarre. They can produce up to five different spore types and are exquisitely fussy about which species of plants they will infect. The most striking oddity of some Rust Fungi is that they need two unrelated hosts to complete their life cycle, so your pear trees will be completely safe from *G. sabinae* unless there is a juniper growing near by.

I will not try your patience by relaying the baroque progression from one spore to the next but merely note that, in order of appearance, they are called spermagonia, aeciospores, urediniospores, teliospores and basidiospores. The last of these are what most mushrooms and toadstools produce on their gills: something that is occasionally seen as a brown, dusty mark deposited by a mushroom left on a kitchen worktop. Many species of Rust form all five spore types, sometimes on just one host, sometimes requiring two. In the latter case only the last three spore types are formed on the principal host, the others being on the alternative host. Such plant pairings are extremely odd: pear trees and juniper I have already mentioned, but there are also pairings of barberry and certain cereals such as buckthorn and oats, plus a handful more.

Rust Fungi are extremely important, but not at all in a good way. They have been a literal pestilence on humankind for millennia, destroying crops and causing devastating famines. Their particular fondness for grasses is the primary reason for this, as most cereal crops are at risk, with only rice being unaffected. With an unhappy relationship with humans going back thousands of years, it is no surprise that efforts have been made to ameliorate the deadly effects of Rust Fungi.

The most famous (and least successful) of these was the Roman festival of Robigalia. On 25 April each year a ceremony would take place to propitiate the agricultural god, Robigus. At some point it involved the sacrificial offering of the entrails and blood of an unweaned red puppy. This is all very unpleasant, but no doubt it could have been far worse. It is thought that Robigalia is the forerunner of the feast of Rogation in the Christian Church, which also takes place on or around 25 April, though no puppies are sacrificed.

We have become much better at propitiating troublesome gods over the years, invoking fungicides, selective breeding of resistant strains of crops and more careful management to keep losses to a minimum. With species that require two hosts to complete their life cycle it is only necessary to remove one host or replace either host with a resistant strain to eliminate the disease. Thus, for example, the disease known as 'white pine blister rust', which is caused by the Rust Fungus *Cronartium ribicola*, spends part of its time on various *Ribes* species (currants and gooseberries), so removing, or changing to immune strains, all the currants in the area will (or at least might) prevent the infection and subsequent destruction of commercially important pines.

Slime Moulds

> *O, what is it, proud slime will not believe*
> *Of his own worth, to hear it equal praised*
> *Thus with the gods!*
> Ben Jonson, *Sejanus: His Fall*

In a suburb of Dallas, Texas, in 1973, residents became concerned about a mass of pulsating, growing and slowly moving yellow blobs that had taken to wandering around local lawns. Efforts to disperse them using a hose just spread them about, turning them into *more* pulsating, growing and slowly moving yellow blobs. Reported in newspapers as the 'Texas Blob', there was, not unreasonably in my opinion, concern that it was an alien life form that was determined to take over the planet, or possibly a mutant bacterium with the same intent. Fortunately, a couple of local botanists helped everyone get some sleep by reassuring them that it was a very common Slime Mould that was completely harmless, lacking in imperial ambition, and that it would disappear quickly, which it did.

No doubt several of the Texan locals thought it was some sort of fungus, and fifty years ago most experts would have agreed. But things move on, and Slime Moulds are now known as being in an entirely different group of organisms, though researchers have not entirely settled on which, with the 'Protista' leading the field. In fact, there are several types of Slime Mould, this one belonging to the Myxogastria and generally termed a 'plasmodial or acellular slime mould'. These, by far, are the type most seen – the rest being small or microscopic, and even translucent. The whole group was, until recently, known as the Myxomycetes, but it covered five orders (a fairly high taxonomic rank), not all of which were closely related to the others. 'Myxogastria' means 'slime-stomach' and is one of two orders that are thought to be closely related, the other one being the cellular Slime Moulds, the Dictyostelia. Together they are classed under the Mycetozoa, which translates as 'fungus-animal', even though they are neither fungus nor animal. Even with all these painfully complex groupings, there are not many species of Slime Mould within the five orders: fewer than a thousand.

I see the Texas Blob (more respectfully known as *Fuligo septica*) every year, always in the summer or early autumn, when Slime Moulds make an appearance, usually creeping around over leaves and fallen logs in the woods. It is indeed a brilliant yellow and sometimes grows to several centimetres across, before crusting over with a greyish layer above a mass of dark spores.

More common and, I think, more spectacular, is *Mucilago crustacea*, which grows and moves about on grass, though its movement is slow and it is always possible to run away from one. The photograph shown here was taken in the mild October of 2020, and shows one of about fifty specimens growing on an old meadow. It is quite extraordinary and looks, in this instance, remarkably like a particularly well-made scrambled egg – it is more usually white to pale cream. I have seen single patches stretch to a metre and more. Unlike most Slime Moulds, this one has a common name.

One of the many great enthusiasts I have encountered over the years is Professor Bruce Ing. I have met him on several British Mycological Society forays and, though he is an accomplished mycologist, his professional interest is in Slime Moulds. On one wander around wood and field we came across *Mucilago crustacea*. Most of our company knew what it was, but Bruce called it the Dog-Sick Fungus on account of its appearance. He would joke that it is a very interesting species and worth collecting a specimen of to examine under the microscope, but that one should be careful because sometimes it *is* dog-sick. This name has stuck, and I believe that Bruce is its source.

Like so many groups of organisms, there are some Slime Moulds that you never see, some that you see occasionally and some that you see all the time. One common species that always causes comment and puzzlement in equal measure is *Enteridium lycoperdon*, sometimes called the 'False Puffball'. It looks very

Mucilago crustacea

much like a small (4 centimetre) puffball fungus except that it is always found stuck to a tree. When it first appears, it is very soft and wet, but it soon firms up and acquires a grey/white skin. Break one open and densely packed brown spores are revealed, looking just like compressed cocoa powder. There is a similar, much smaller and decidedly pink/peach Slime Mould called *Lycogala terrestre*, which normally grows in small groups. It too is grey and full of cocoa when mature, but while still young the broken skin will reveal a brightly coloured and slightly sticky fluid.

Many Slime Moulds form minute works of abstract art. Imagine a 1-centimetre-diameter patch of rotting wood which contains fifty tiny and brilliantly coloured (purple is common) upright lollipops, all clustered together. What are these things, how do they grow, how do they *move*?

First of all, a small but shameful confession: my one total failure in my O-level exams (GCSE in modern British terms)

The 'fruiting bodies' of a Slime Mould, possibly *Trichia decipiens*

was biology. My tutor, Mr Kent, was an affable old fellow, but seemingly too bored and set in his ways for inspiration to be a part of the curriculum. With few textbooks available, we relied on him scribbling words and drawings on the blackboard and us copying them down. One of the drawings was of an amoeba. I dutifully copied it, and in revising amoebas for the exam, I copied it again and again, resulting in a Chinese whisper of what should have been and an ignominious fail on the big day.

Perhaps you will forgive me an irrelevant aside, which might at least show you what we were up against. The taciturn Mr Kent (known inevitably and ironically as 'Superman') bore sole responsibility for the sex education of us grammar school boys. He had not a word to say on the subject but, moving on from amoebas, or maybe the earthworm, to 'Human Biology', subsection 'Reproduction', simply wrote the following on the blackboard: 'Sexual congress is undesirable prior to marriage.' Eight-stone vats of hormones that we were, we knew with painful certainty that this was untrue and never trusted him again.

Despite my inability to provide even a vaguely accurate drawing of one, I have always found amoebas fascinating – all that extending bits of themselves and the fact that you can see straight through them. What I did not expect was that they would rear their tentacles on fungus forays.

Mr Kent told us that amoebas were primitive animals – after all, they do move about – and they often are animals, but amoebas, or at least 'amoeboids', appear at one stage or another of the life cycle in all the major groupings of cellular life. Here is how they occur in Slime Moulds.

Unlike the spores of fungi, which produce a hyphal tube when they germinate, Slime Mould spores split open to produce one to four cells known as protoplasts. Each of these contains a single set of chromosomes (i.e., is haploid). These cells will either behave like amoebas, complete with tentacles (less fancifully known as pseudopods) or they will acquire two thin flagella (think mutant tadpole). The amoeboid form is employed in relatively dry conditions, the flagellate form if there is surface water, and they can change from one mode of propulsion to the other as circumstances dictate. The amoeba form is called a myxamoeba and the flagellate form 'swarm cells', the latter term being straight out of a sci-fi movie. These protoplasts feed by ingesting bacteria, and they 'breed' asexually.

At some point compatible protoplasts fuse completely to form a zygote, which also moves using either amoeboid or flagellate forms. The zygotes themselves reproduce by division, until there are countless numbers all together to form a single, gigantic cell. This is called a plasmodium, and it was several plasmodia that were wandering about that Texan yard.

Everything about Slime Moulds is extraordinary, but the plasmodium defies any expectations. It tends to grow by fanning out in several directions, dying off where food is in short supply and congregating where it is not; hence its ability to move. As it grows, thread-like structures trail from the leading edge. These are effectively vascular (vein-like) in nature, their 'wall' consisting of, well, slime. The fluid and zygotes within the tubes are pumped by peristaltic action, directed by calcium ion concentrations: that is, they are pumped towards an area where calcium concentrations are high. Being peristaltic, the movement is a slow pulse.

Once it has eaten everything in its path, the complex 'fruiting' bodies are formed, and it is these, plus the larger of the plasmodia, which we most often encounter on our walks. At least you can now enjoy the bright beauty of Slime Moulds without feeling the need to call in the air force.

Stinkhorn

Should I omit or recount your shame, red Priapus?
Ovid, *Fasti*, vi

This is so common a fungus that I am surprised when people tell me they have never seen one before. To be fair, they often have a little trouble telling me *anything* as they are usually too busy giggling. While, to the initiated, this is just another fungus, albeit an amusing one, to most people the Stinkhorn is a mystery, and it is with this paper-thin justification that I have recounted its shame in the book. The Stinkhorn has the Latin name it thoroughly deserves: *Phallus impudicus*. This means 'shameless penis', though 'enthusiastic penis' seems a touch more appropriate. Difficult as this bawdiness is for some people, we must at least be thankful that we have not retained the common name conferred by the sixteenth-century botanist John Gerard, who called it the 'pricke mushroom'.

The Stinkhorn is a leaf-litter fungus found in most types of woodland and growing on whatever organic detritus has accumulated on the forest floor. The immature stage is known as the 'witch's egg'. It is the shape and colour of a large ping-pong ball and attached to its underground mycelial mass

Two flies have been inspired by a Stinkhorn

by a white rhizoidal cord. Sometimes these eggs are mistaken for puffballs, but they are two or three times as heavy for their size. Cut in half vertically (as seen on p. 8), the structure of the fully developed Stinkhorn can be seen in highly compressed form surrounded by a mass of jelly, the whole covered in a thin, leathery skin. As the compressed mass develops, it bursts through the jelly and grows upwards to its full height of about 15 centimetres. It does this very quickly, but reports of 10–15 centimetres per hour sound more like 'personal best' claims rather than the 3–5 centimetres per hour with which I am more familiar.

The white skin and jelly mass (peridium) remains as a base to support the 'stem'. The stem is the shape of a hollow cigar (still pointy at both ends) and made of an expanded material that I can only describe as 'prawn cracker' (it even crunches when you squeeze it!). On top of the stem there is a 'cap' known as a gleba. This is barely attached to the stem, like an otherwise empty and upright bell balanced on a finger. The outside of the cap is reticulated like a honeycomb and covered with a dark green sticky substance that smells of rotting meat. You will find it to be very sweet should you be mad enough to taste it, as I once did. It is the gleba that holds the spores. Flies are attracted to this substance and feast on it. The flies are generally species of carrion fly, including the curiously named *Calliphora vomitoria* (roughly translated as 'Puking Beauty'), the Bluebottle. When they fly away, they will deposit the spores in the manner to be expected. In an unpleasant twist to this otherwise delightful story, the material that makes up the gleba is believed to contain a laxative. This ensures that the flies do not hold on to the spores for too long, depositing them while still in the forest. If you thought that this explains the fly's name, despite it involving the wrong end of the fly, I must tell you that it does not. *Vomitoria* is a reference to the fly's common habit of regurgitating enzymes to pre-digest its food.

A Stinkhorn cap will be completely cleaned of its sticky green spore mass in a few hours, the spores distributed for hundreds of yards. In its entirety, it is a highly sophisticated and effective method of spore dispersal. It does not rely on the vagaries of wind dispersal used with most larger fungi, and its mission is accomplished sufficiently quickly to avoid accidents or being eaten.

I have seen these things hundreds of times and they always engender a smile. I used to run local government adult education fungus forays on autumnal Saturdays. We would look for fungi in the morning and identify them back at the classroom in the afternoon. On several occasions we found immature Stinkhorns, and after the class I would place them in a saucer, nestling in some wet tissue paper, and take them to the office where the two women who handled the admin worked. By Monday morning the office stank like an abandoned abattoir. It was hilarious, but a great pity that some people have no sense of humour.

The cocktail of chemicals used to attract the flies is interesting, not least in that it attracts only female flies. This is, obviously, because only the females lay the eggs, and they do so on carrion so that there is food for their offspring (maggots). There is no reason, therefore, for males to find the smell of carrion attractive. Twenty-two volatiles have been discovered in the mature Stinkhorn; most of them are aromatic and most of these unpleasant-smelling sulphur compounds. Methyl mercaptan is the primary smell you notice when you cut a mammal in half from top to tail; dimethyl trisulphide smells of rotting onions; hydrogen sulphide is familiar as rotten eggs. But there are some pleasant smells as well. Dimethyl disulphide is the dominant smell when you open a can of sweetcorn and the primary aromatic of truffles. Phenylacetaldehyde has hints of roses and hyacinths; linalool is somewhere between mint and rosewood; and ocimenes are fruity with a touch of basil. Unfortunately, none of the pleasant aromas can make up for the horror of the unpleasant ones.

The fragrance of the Stinkhorn can be detected from thirty paces, but it is not the only British member of the Phallales (the order that encompasses these oddities) determined to make parts of the countryside no-go areas. Among them is the rare *Phallus hadriani*. It may come as a surprise, but coastal sand dunes are excellent places to find fungi, with perhaps a dozen species found only there. *P. hadriani* is one of them. It is very similar to the Stinkhorn, but easily distinguished by its habitat and pink/brown egg stage.

One more similarly phallus-shaped species is found in Britain, the Dog Stinkhorn, *Mutinus caninus*. The generic name comes from a Roman god of marriage, Mutunus, though precise details of how he was worshipped will not be revealed here. The references *caninus* and 'dog' simply refer to the smell they produce – that of dog's poo. I have not inquired what chemicals are employed, but I suspect mercaptans are in there somewhere. The Dog Stinkhorn is much smaller and thinner, and has an orange stem and a more pointed cap.

Moving away from penises for a moment, there are two members of the Phallales occurring in Britain that don't look like penises at all. They are both from warmer climes. They are the Devil's Fingers, *Clathrus archeri*, which arrived in Europe 100 years ago from Australasia, and the slightly more common Cage Fungus, *C. ruber*, from continental Europe and a more recent British acquisition. In keeping with their origins, they have been found mostly in the southern counties of England. Both are quite astounding in appearance: they burst out of an egg, the latter to form a distinct cage structure, the former opening like a hand, fingers arching backwards. Both are a vivid blood red in colour, their inside surfaces coated with the dark green gleba and smelling of rotting flesh.

I cannot leave the subject of the Phallales without mentioning the famous story of Etty Darwin. She was the daughter of Charles Darwin and aunt of the wood-engraver Gwen Raverat, whose memoir *Period Piece* was published in 1952. In it she recalls Aunt Etty wandering around the woods with a basket and pointed stick, sniffing out Stinkhorns. When she found one, she would impale it and drop it into her basket. Raverat describes their subsequent fate: 'At the end of the day's sport the catch was brought back and burnt in the deepest secrecy on the drawing room fire with the door locked – because of the morals of the maids.'

Cramp Balls

And the earth shook and the King stood still
Under the greenwood bough,
And the smoking cake lay at his feet
And the blow was on his brow.
    G. K. Chesterton, *The Ballad of the White Horse*

During a pit-stop in the New Forest, on my way back from a Cornish holiday, I found my first Cramp Balls; not that I knew what they were at the time. Hemispherical, about 40 millimetres in diameter, jet black, rock-hard lumps and attached to a tree, they looked like nothing I had ever seen before. They appeared to be completely dead, and when I pulled one away, I found it to be light in weight and crumbly – like charcoal. Nothing in my understanding could explain how it got there, and I was suitably annoyed at my ignorance. But I am indebted to those Cramp Balls for they were my inspiration for a lifetime's interest in fungi.

In the 1960s, if you wanted to know what anything was, you either asked someone who might have an idea or went to the library. I went to the library. A quick search through the diminishingly small number of books on fungi found it to be the Cramp Ball, *Daldinia concentrica*. I learned that it was a common fungus of dead ash and occasionally other trees. It grows, a layer at a time, forming the concentric rings that were visible in my broken specimen. I was completely content with what I had discovered and left it at that, except that I spent five shillings on *The Observer Book of Fungi* (I still have it) and kept my eyes open for other fungi.

Still, I learned a little more over the years: that the name was a reference to their imagined ability to deter cramps (they don't), and that there is another name for them: King Alfred's Cakes, though not, as my young daughter once suggested, King Alfred's Balls. I also discovered their ability (when dry and broken) to smoulder gently for twenty minutes after being lit by a spark from a fire iron; it is one of the tinder fungi. The red glow is only

Cramp Balls – one cut to show concentric growth-layers

visible if you blow on them gently, and I doubt if I am the only one to have put a smouldering Cramp Ball in my pocket thinking it had gone out, only to find out that it hadn't when my coat caught fire. However, there is much more to this fungus than silly names and party tricks.

First of all, there are twenty-eight species of *Daldinia* known across the world, six of which occur in Britain. *D. concentrica* is extremely common in this country, while the other five are rare or very rare. I have only seen one other species, *D. fissa*, growing on some burned gorse in Hexham, Northumberland. *Daldinia* as a group are considered to be 'xerophytes', which means they are adapted (or at least that their ancestor was adapted) to dry conditions. This neatly explains *D. fissa* insisting on burned gorse, its only habitat. Many species occur in the drier parts of Mexico, another indication that *Daldinia* are adapted to low moisture levels.

Daldinia species are also 'endophytes', harmlessly taking up residence inside their host tree or shrub and biding their time until the tree, or part of the tree, starts to die and stops producing sap. They then burst into action, filling the available dead tree with mycelial threads, which consume both cellulose and lignin and eventually form their familiar fruiting bodies.

D. concentrica takes its name from Agostino Daldini, a nineteenth-century Swiss monk and a friend of the mycologist Vincenzo Cesati, who named it. '*Concentrica*' is a morphological name and an obvious reference to the conspicuous concentric rings that can be seen inside broken specimens. These look very much like tree rings but are not annual; instead they are rapidly laid down in a few months. To my shame, I always thought that each layer was fertile and produced spores, but it is not so. Only the final layer produces spores. My apologies to the 5,000 or so people I have misdirected over the years on my fungus forays. What, then, are the other layers for? They are made of fibre and tube, laid down with an alternate concentric and radial orientation and full of complex plumbing. The hypothesis is that it is a water-conservation strategy, likely in something that is arid-adapted.

Daldinia species are 'ascomycetes', meaning they belong to one of the large divisions of fungi, the Ascomycota. Most of the familiar fungi that one sees in woods and fields are in another division, the Basidiomycota. But there are some very familiar ascomycetes, such as truffles, morels, the black mould that infests damp houses (and sometimes people's ears), *Aspergillus niger*, and, more pleasantly, yeast.

Ascomycetes are informally called the 'spore shooters' because their spores are shot from the end of sausage-shaped bags (asci) in which they are formed. Asci usually contain eight spores, and it would be better to say 'squirted', as the asci are filled with slimy water at high pressure and the spores are squirted out through a trapdoor at the top when the pressure is sufficiently high. This is most easily seen on a remarkable fungus called the Orange Peel fungus, *Aleuria aurantia*, which, if disturbed, will release millions of spores all at once to form a noticeable puff of 'smoke'. These have been ejected from the asci, which are all standing on end, shoulder to shoulder, within the bright orange upper surface.

Daldinia is more discreet, inconspicuously firing its spores from tiny holes covering the black surface. Employing one of the most remarkable microscopic mechanisms to be found in the living world, a single hole (ostiole) can shoot out millions of spores. Beneath each ostiole is a flask-shaped structure called a perithecium, which is cylindrical and more or less tapering top and bottom. *Daldinia* species belong to a group of fungi known as 'flask fungi' because of this structure. Inside the flask there will be thousands of asci and the potential for many more. Like a machine gun firing bursts of eight bullets, each one will grow up to the ostiole, release it spores and then wither and withdraw. It is then replaced by another. It has been estimated that a single perithecium can produce 100 million spores in a single day. They are only about half a millimetre in diameter and tightly packed together, so a single Cramp Ball will host thousands of them. If this was not sufficiently impressive, *Daldinia* species have long active lives, mass-producing spores from May to December.

Cramp Balls can even produce spores for several weeks after they have been detached from their host tree. Should you come across a fresh specimen, it is worth taking it home. It needs to be mature and active, something easily spotted if you rub the surface with a finger – it will come away blackened by spores. Place your specimen in the middle of a large sheet of white paper, cover it with the largest bowl you can find and leave it overnight. The next morning you will see a corona of black spores shot out from the Cramp Ball. Should you repeat the experiment during the day, very few spores will be seen, as *Daldinia* species practise what has been evocatively and carelessly termed 'nocturnal ejection'. The hypothesis here is that spores do not stick to the dry daytime surfaces of host trees that would occur in the arid habitats of *Daldinia*'s progenitors. However, night-time dew dampens the surface of any potential host tree that the spores land on, enabling them to stick and germinate.

One would imagine that the above-mentioned production and dispersal of spores would be the end of *Daldinia*'s reproductive repertoire. But ascomycetes such as this are hugely complex and a great trial for the novice trying to understand them. They almost invariably produce asexual spores as well, contrasting with the sexual spores mentioned above.

The asexual phase is seldom seen but is known to grow sometimes on the sites of old Cramp Balls (the sexual phase).

The asexual phase consists of bodies that are soft, white and wetly furry. The spores that they produce are known as 'conidia'. Spore production with these asexual fruiting bodies is achieved by the slightly primitive method of 'snipping off' short lengths of hyphae which contain haploid (single chromosome) nuclei. Asexual pores can also be made on structures known as conidiophores.

The sexual spores are those produced by the fruiting body we know as the Cramp Ball. Mating occurs between strands of hyphae within the wood. The hyphae produced from this mating will join with a much larger mass of 'unmated' hyphae to produce the fruiting body, the bulk of the body being made by and of the latter hyphal type.

Green Cup Fungus

A livelier emerald twinkles in the grass
Alfred, Lord Tennyson, *Maud*

Green wood is an unexpected find in woodland; and by green I do not mean that it is unseasoned, just *green*. Anyone wandering around oak, beech or hazel woodland is very likely to find small pieces of wood lying on the forest floor that display a bright flash of green. If it is examined carefully and the loose, rotten surface rubbed away, the remaining wood is usually *entirely* green. Apart from being so striking, this is a rare reminder of how much room is taken up by a fungal mycelium in any infected piece of wood: all of it.

While the green, mycelia-packed wood is a common occurrence, the fruiting bodies are seen less frequently. I normally find them once or twice a year, and what a joy it always

is. While the wood is a startling blue/green, the tiny cup fungus that grows from it is an electric verdigris. Sometimes there will be hundreds on a single log. The colour comes from a chemical called xylindein, a word that means 'indigo wood'.

There are two species that turn wood greenish and produce verdigris cups, and they are impossible to distinguish without examining the spores under a microscope. Unfortunately, the two accompanying names are also (almost) identical, causing great confusion. They are *Chlorociboria aeruginascens* and *C. aeruginosa*, the latter being rare and with much smaller spores. Latin names and mycology in general are difficult enough for people without these self-inflicted confusions.

Oak infected with the Green Cup Fungus was once used in Tunbridge Ware. Tiny squares of different-coloured woods are arranged to form patterns or pictures on such things as small boxes. The green/blue wood was just what was needed for leaves, seascapes and, to a lesser extent, the sky. I have seldom found any green wood strong enough to be used for anything, so am at a loss to know their source.

Hypomyces

Hyt is not al golde that glareth.
  Geoffrey Chaucer, *The House of Fame*

I have been reasonably restrained in not including too many fungi, but two members of the genus *Hypomyce* were always going to make it. They are *Hypomyces chrysospermus* and *H. lateritius*.

One of the most striking of all mycological sights is a Bolete infected with *H. chrysospermus*, recently dubbed the 'Bolete Eater'. Boletes, if you did not know, are mushrooms with tubes underneath the cap instead of gills. The very edible Penny Bun, *Boletus edulis*, is one of them. The parasitic *H. chrysospermus* fungus can be almost blindingly yellow, upstaging even the Fly Agaric for sheer presence. I see it several times every autumn, so it is extremely common. As the common name indicates, it is found on fruiting bodies of members of the Boletes, though it does not seem to be particularly selective of species. It coats the surface with a white, then powdery brilliant yellow, then brown skin, while disintegrating the inside of the host mushroom into a vile, sticky and smelly mess. The shape of the mushroom becomes deformed and barely identifiable as a mushroom at all.

The various coloured skins indicate the life stage of the parasite. The white and the yellow are anamorphic stages, producing asexual spores, and sexual spores are produced by distinct ascophores during the final, brown stage. The latter are visible with a good lens, but few will wish to get that close to one by then. The yellow seen on one of the asexual stages is that of the spores. These get everywhere when they have the chance and provide part of the species' Latin name, *chryso*, meaning 'golden'.

It was some years after I first saw the second of the two *Hypomyces* species that I discovered what it was. It was when I read about the 'Lobster Mushroom' that all became clear. The North American Lobster Mushroom is a fungus – two fungi, in fact – which form a bright lobster-coloured structure, vaguely in the shape of whichever species of Milkcap fungus (*Lactarius* species) has been parasitised. Mushroom hunters are all too familiar with edible fungi falling into an unstructured mess when

they are attacked by a parasitic fungus, so it is quite unexpected to find them transformed into more robust versions of their old selves. But this is what happens with Milkcaps when they are parasitised by *Hypomyces lactifluorum* (see image on p. 228).

This is a precise parallel with what I had seen with *Lactarius deliciosus*, the Saffron Milkcap. As with the Lobster Mushroom, the otherwise delicate mature fruiting bodies were very firm in texture, and the gills underneath the cap were no longer visible, replaced by shallow, pale grey wrinkles, though the rest of the mushroom retained its shape and colour.

Lobster Mushrooms, despite their unfortunate status of 'rotten fungus', are edible and highly prized, complete with recipes in North American cookbooks. I presume that an infected Saffron Milkcap would make excellent eating too, but I did not get to be so old by presuming.

Nostoc commune

What a monstrous and abnormal Lichen thallus comes into view.
Proceedings of the Royal Irish Academy, 1870

There are few organisms in the British countryside for which the first reaction from people is 'Yuck. What's that?!', but *Nostoc commune* is one. It goes by several viscous names: witch's butter, mare's eggs and star jelly are among the common names in use, though none is satisfactory, and they are sometimes applied to other 'yuck' species. For our purposes we will stick with the Latin name, *Nostoc commune*. It is frequently referred to as an alga, but that is not what it is: *Nostoc* species are cyanobacteria.

We do not normally expect bacteria to appear as a vast swathe of wrinkled, glistening, green/brown jelly-like blobs, covering gravelly footpaths and making unwanted guest appearances on badly tended lawns. What we do expect is for them to stay invisible: bacteria are, after all, microscopic. The cyanobacteria are a division of the domain Bacteria, which includes twenty-nine other divisions. Some cyanobacteria may be familiar: 'algal' blooms, which effectively poison (in turn) seawater, shellfish and us; and 'spirulina', a foodstuff/additive made from the cyanobacterium *Arthrospira platensis*. The bright colours that are sometimes seen in the Red Sea and in some other nutrient-poor waters are due to the halophytic cyanobacteria that thrive there. Closer to home, you may see cyanobacteria every day, as it is these and/or true algae that provide the photosynthetic component of lichens. Many of these are other *Nostoc* species, and the one pictured on the next page was in a small, shallow depression on a tomb in the local churchyard.

How is it, then, that a microscopic bacterium can grow so big? Obviously, it is partly because a single blob of the jelly is made out of billions of individual bacteria, a colony, so then we must ask how it is they form a jelly. In *Nostoc commune*, individual, rounded, photosynthesising cells, interspersed occasionally with larger nitrogen-fixing cells, join to form long chains which are covered by a mucilaginous sheath. The cells exude a mix of proteins and polysaccharides for protection, and to hold everything together. They start as a single blob, then expand to larger blobs or discs or wrinkly discs.

Several other species exist, one of them being *Nostoc pruneforme*, the specific epithet meaning 'plum-shaped', which indeed it is, complete with a hollow where the stone would be. *Nostoc* appears to be a word conjured up by the sixteenth-

The cyanobacterial lichen *Collema auriforme*

century alchemist Paracelsus. He combined the old English and the German words for 'nostril', *nosthryl* and *nasenloch*, a reference to something that is notorious for producing extracellular polysaccharides, this time in the form of snot. Snot or not, these are very tough organisms, surviving dry weather and able to reconstitute completely when it rains again.

Witch's Broom

But aunt Madge had a witch's broom, to sweep cobwebs out of the sky.
Sophie May, *Little Folks Astray*

One of the great delights of taking sufficient interest in things to be able to spot and identify them from a distance is the chance it affords to inform anyone within earshot as to what you have seen. Such comments as 'You point out those bloody Wild Horseradishes every time we drive by them' make it all seem worthwhile. Identifying mushrooms (accurately or not) while driving past them at 50 mph is another of my tricks. As is spotting things in trees. 'Oh, look, there's a rookery, no, they're those horrible Crown Galls. Wait! Witch's Brooms!' I should have got it right first time. Witch's Brooms have a central, woody mass with multiple thin shoots sticking out, and are unmistakable.

Galls in general are dealt with elsewhere in this book (p. 138), but Witch's Broom, which is also a gall, is sufficiently conspicuous and common to warrant a chapter of its own. This is one 'mystery' that will not require a walk around the countryside as it is commonly found on roadside trees and in parks and gardens. However, it can be confused with several other masses of twigs seen in trees, such as squirrel dreys, bird's nests and certain other galls.

Witch's Brooms occur on several tree species and are generated by a number of galling agents: mites, bacteria, viruses and fungi. In Britain they are nearly always seen on birch trees and nearly always caused by the fungus *Taphrina betulina*. *Taphrina* is a career genus, its vocation being that of parasitising plants, as with *T. pruni*, which infects *Prunus* species. It is worth mentioning that, confusingly, there is another *Taphrina* species that grows on birch, *T. betulae*. This causes only 1-centimetre discs on otherwise healthy leaves.

Witch's Brooms can be a spectacular sight, dominating leafless winter birches. Most of what you can see is a mass of shoots and maybe a central mass, but sometimes they will produce roots

which are doomed to hang metres above the ground. The initial growth is from an infected axillary bud (from between a branch and a leaf stem). The growth is swollen at the base and extends to form a sizeable branch, followed by a multitude of thin shoots, long in the first year of growth, shorter in subsequent years. The shoots frequently die, leaving an aerial bush of living and dead material.

In summer these can be invisible, hidden by the normal leaves. They can live for as long as their host tree, and more mature specimens will develop a central, solid and irregularly spherical woody mass. It is on deformed leaves growing on short shoots that the fruiting bodies are formed. As with all other gallers, the fungus takes over control of the host plant's growth, but considering how discreet most galls are, *T. betulina* is like a rock band trashing a hotel room. While most authorities consider it to be fairly benign, it is a problem for timber production, reducing timber yields by 25 per cent. On the plus side, the mass of twigs forms an excellent habitat for several invertebrates.

Hair Ice

The great in Honor are not always wise:
Nor Judgement under silver Tresses lies.
  George Sandys, *A Paraphrase upon the Divine Poems*, 1638

I have never seen Hair Ice. Some of my friends in Scotland have, and I am jealous, but they have the advantage that it is cold up there and for longer, and Hair Ice obviously requires cold weather. It is quite beautiful, and an arresting sight by all accounts: thousands of 0.02-millimetre-diameter strands of ice anything up to 20 centimetre long, growing on a log and creating an unusual and extravagant hairdo. The hair is a silver-white, of course, but still luxuriant and in good, shiny condition. The hairs are rooted on the surface of dead, broadleaved tree branches that have lost their bark and are usually lying on the forest floor. Although the temperature at which Hair Ice forms needs to be a little below 0°C, the wood itself must not be frozen.

Considering how very delicate these structures are, they can last a long time – hours or even days if the weather is amenable. Such stability comes as a surprise, as the small crystals of which

it is composed should rapidly recrystallise into larger crystals at such relatively high temperatures, and the entire structure fall apart.

Something very odd is happening here, but how to explain it? Various theories have arisen over the years, though none proffered with great confidence or, at least, detail until recently. Nevertheless, a fungal basis to the phenomenon had long been considered likely and was confirmed, complete with an eye-watering amount of detail, by a German and Swiss team in 2015.

They showed that Hair Ice only occurs in the presence of the fungus *Exidiopsis effusa*. This fungus is as unprepossessing as they come, invisible inside a piece of wood for most of the time and producing only a delicate white coating to the wood surface when it fruits. It does not need to be fruiting to form Hair Ice, but it must be in the wood and alive. The hairs themselves are created from water being sucked (their word) from the pores of radial xylem vascular tissue (tubes). It freezes at the surface, forming microscopic ice crystals. Without the fungus present, ice just accumulates as an untidy layer of crystals, so the fungus must have a role in forming the hairs. They note that, while melted strands just collapse, those that happen to be attached at both ends will leave a mass of water drops suspended on a micro-thread. Also, that the melted water from Hair Ice is brown in colour. With something that is obviously not water producing a thread and the discoloration in the meltwater, they hypothesise that one of the chemicals created in the breakdown of wood lignin by the fungus acts as an antifreeze, preventing the recrystallisation of the microscopic ice crystals. I am leaving a lot out here, so do read the paper if you are interested and have four hours of your life going spare.

Elm Wings

[…] the War of the Gods, the Winged Tree, and the Veil
  Algernon Herbert, *Nimrod: A Discourse on Certain Passages of History and Fable*, 1828

Sometimes one has to admit defeat. Over the years I have noticed, if only infrequently, that some young tree branches have layered, corky 'wings' on them. I found them again in 2020, and you can see my find pictured here. It is on an English Elm, *Ulmus procera*, and, incidentally, besides the wings, you can see some galls on the leaves. I asked around and did a little research. A few botanically inclined friends thought it was normal for elms to have these wings; other research declared them to be the primary symptom of wingbark virus or winged cork virus, though there were no research papers or articles about such a virus that I could find. In addition, several species of elm, notably the Winged Elm, *Ulmus alata*, come with wings fitted as standard. Why it appears on English Elms only occasionally, or why it only affects some

small branches on a single tree, I do not know. Perhaps the evident predilection for elms to produces these wings is expressed under certain conditions; perhaps that condition is a virus. I do not know, but someone must. For now, I will just admire them for the pretty puzzle with which they have presented me.

The Seashore

Oyster Beds

Poor Britons, there is some good in them after all – they produce an oyster.

Sallust

I don't see native oysters, *Ostrea edulis*, very often, either on my seashore perambulations or on a plate. The determined, or well-situated, longshore loafer might encounter the odd one or two at an exceptional low tide, but that is all. Much more often I find the native oyster's replacement, the Pacific oyster, *Crassostrea gigas*. It occasionally occurs in vast numbers, cemented to a sea wall and sometimes to the beach itself, forming a living, indeed *edible* pavement.

Native oysters are still landed in British waters, but no longer on the scale of previous centuries, when they were famously cheap. During the first half of the twentieth century the British catch fell to almost nothing owing to a conspiracy of disasters: overfishing, with areas of seabed dredged bare, disease in the form of single-celled parasites such as *Bonamia ostreae*, competition from the wonderfully named Slipper Limpet, *Crepidula fornicata*, and sewage pollution.

The more robust Pacific oyster, an animal that is now farmed on an appreciable scale, was introduced to British waters in 1965. It is generally farmed, or at least 'fattened', in cages which are given the occasional vigorous shake to prevent them cementing themselves into one large lump. The ones we buy are still very young, but Pacific oysters can grow to an enormous size if given the chance. Years ago someone sailed or powered through an oyster farm in a local harbour, breaking open cages and scattering thousands of young oysters all over the seabed. They continued to grow, and every now and then a few would be collected. Simon, a friend of mine who used to run a plant that purified shellfish ready for market, would receive the occasional consignment, which he sold as XLs. Some of these should have been XXXLs: the largest I bought from him came in at just under 2 kilos, though there was one nailed to the factory wall that was nearly a foot long. They were impossible to open in the usual way, so I had to cut through the adductor muscle with a thin, sharp knife. The chicken-breast-sized flesh inside needed to be chopped up and cooked because only the exceptionally talented could swallow one whole.

Oysters have long been encouraged by man, with nursery beds created to breed and grow the young animal. Once large enough, they were taken to fattening areas offshore to grow to their full size. Beds were also used to store oysters before dispatching them to market. Many of these nurseries are still operating, some of them on old or even ancient sites, but most are just ruins and shadows, their existence revealed by a series of rectangular banks close to and often in the sea. There may also be an abandoned jetty in an odd-looking place which once served the oyster farm, or simply a track through the salt marsh to where one once existed. It is still quite possible simply to come across them on a visit to the seaside, but for a systematic search there is aerial photography.

Oyster farming was, for the most part, confined to the south and eastern coasts, Essex, Kent, Sussex and Hampshire being the strongholds of former ostreiculture. In the convoluted tidal channels that make up much of the Essex coast there was once a large and thriving native oyster industry. Scores of abandoned nurseries can be seen along the Rivers Roach and Crouch, and there are more in the Medway. One on the River Roach is still in operation. The furthest north that I can find, in England at least, is at Alnmouth in Northumberland.

Just to the east of Portsmouth lies Langstone Harbour, a place I mention within these pages for its salterns (see p. 212). To the east of this is Hayling Island, from which one can see the coastal town of Emsworth. Either side of the north of the island there are two well-preserved Oyster Beds. Oyster cultivation in this area goes back a very long way, at least to William the Conqueror, when it was recorded in Domesday. Writing of the Hampshire harbours in 1826, the historian Richard Scott makes it clear that these were once famous:

> *The harbours are some of the finest nurseries in the kingdom for oysters; large quantities of which, under the designation of 'Emsworth oysters' are sold at good prices many miles distant.*

The Oyster Beds at Hayling Island

The one to the east is situated on what can only be described as a vegetated mudflat, known as Fowley Island. There is a tidal track that (mysteriously) *almost* reaches it, so a boat is the only option, not that it is a journey I have ever taken. However, I know the beds to the west very well as they encompass a particularly exhilarating seaside walk. Here are huge banks, curving around for 2 kilometres. The walk can only be accomplished at low tide and still involves a little wellie-work because of the once sluice-gated channels that lead in and out of the lagoons. The entire complex contains seven large enclosures for breeding, five small winter beds and fourteen smaller beds at the southern end where oysters were kept prior to their journey to market.

A few French names have been borrowed for oyster cultivation. The latter beds are called *claires* and the breeding beds are known as *parcs*. To breed oysters, adult specimens are placed in *parcs*, where they will spawn. Mobile oyster larvae feed and develop, eventually settling on and attaching to the various substrata (*fascines*) that have been place in the *parc* for their convenience. Once they are attached, they are known as 'spat'. Much effort and theorising has been put into finding the best *fascines*. Ceramic tiles were among the favourites, stacked up or laid like little roofs. Wooden hurdles were often staked (horizontally) just above the seabed, though thousands would be needed in a large *parc*. Bushy branches fixed to posts, looking like grim and sunken caricatures of trees, were also employed.

Once they are of the correct size, the young oysters (brood) would either be shipped off to oyster-growers in Essex and Kent to be fattened or be moved to a local part of the seabed to which the owner has rights of fishery. In the case of the Emsworth oysters this was a channel in the middle of Langstone Harbour called Crastick's Channel. However, the owner in the early nineteenth century, Matthew Russell, set up home on a small islet in the harbour a little to the north of the channel and it has been known ever since as Russell's Lake, following an inexplicable tradition of calling various parts of the harbour 'lake'. The area was always marked as private fishery with stakes.

The demise of Emsworth oysters was inevitable for the reasons given earlier, but it was hurried rather by a tragedy that occurred in 1902. Oysters, eaten, as they are, raw and indeed *still alive*, are notorious for causing illness. I was chatting to Simon about this one day and asked him if any of his customers had ever been poisoned. Expecting a confident 'No!', I was taken aback by his exclamation of 'Good God, yes!', followed by numerous cheerful anecdotes about cancelled honeymoons, the Great Oyster Disaster of 2001 and so on. None of this, I must point out, would be the fault of the purifying plant as they can only remove harmful bacteria and not viruses, such as norovirus, which are contained within the flesh, not the gut. If you eat enough raw oysters, you will one day succumb to that fast and furious bane of the gourmand.

In the past, no purification process was possible; it was just a matter of hoping for the best. Unfortunately, at a mayoral banquet in Winchester in 1902 given in honour of the former mayor, Mr B. D. Cancellor, only the worst was in attendance. As well as many other delicacies (including an unspecified soup), Emsworth oysters were on the menu. One of the food items evidently disagreed with most of the diners, and it is estimated that most of the guests suffered either 'ptomaine' poisoning (food poisoning), typhoid or, presumably, both. There were fourteen recorded cases of typhus and two deaths from that disease. One of the fatalities was that of a waiter who had evidently helped himself to an oyster or two, and another was a local notable, Dr William England. Although the oysters were seriously in the frame, the caterer complained about rumours that it was his soup at fault and demanded a full inquiry to exonerate him. There followed an enormous fuss, which went on for years, a newspaper report from the time christening the public reaction as 'oysterics'.

By 1905 all was settled. Mr Foster, the owner of the oyster company, which up until then had produced 2 million of the little molluscs a year, sued Warblington Urban District Council for polluting the waters of the Emsworth creek by releasing untreated sewage and unwanted effluent from a slaughterhouse directly into the sea upstream from where some of Mr Foster's beds lay. Despite an unhelpful intervention by a third party to the effect that Foster had *intentionally* situated his oysters downstream of the sewage outlet because there was plenty for them to eat at that point, the judge found for the complainant. Sadly, with even the robust people of Portsmouth now being reluctant to eat oysters, and the various other problems facing the industry, Foster's company never recovered from the blow, and his splendid Oyster Beds are a mere memory.

Salterns

Salt is born of the purest parents: the sun and the sea.
Pythagoras

When I was a child, I used to go to Langstone Harbour, to the east of my former home in Portsmouth. Here I would collect cockles with my family. We would park just off the relatively new Eastern Road, which ran north–south along the harbour, and set off with buckets. On the landward side, going under the road, there was a deep and slightly sinister-looking inlet, controlled by sluice gates. I always wondered what it was for.

Further west and many years later, I was exploring the salt marshes to the south of Lymington in Hampshire. The walking was easy, on well-tended pathways all raised above the marshes. The most striking features of the area were the large and often rectilinear lakes, around which the paths ran. I wondered if it was an early, abandoned fish farm: oysters, perhaps. My opportunity for pleasant speculation was cut short, however, for there was a notice telling the visitor that they were the remains of extensive Salterns. Determined to remove any doubt, it further explained that a Saltern is a place where salt is extracted from seawater.

Salterns once abounded around our coast, and some have been traced back to the Iron Age. Sadly, few have survived so well as those at Lymington. Many ancient Salterns have been lost to rising sea levels, sinking land levels, the occasional catastrophe of a storm surge or the odd tsunami, many to agriculture and most, perhaps, to marinas and other coastal developments. It is not certain, therefore, that you will encounter the remains of one by accident. However, they do sometimes leave traces.

The Salterns at Lymington in Hampshire

Salterns in good condition, such as those at Lymington, will be visible as rectangular pans averaging about the size of a tennis count, and there may be several of them. Sometimes, as (it transpires) at Langstone Harbour, the evidence comes in the form of sluice gates and an accompanying channel. Unnaturally straight and sometimes criss-crossing channels in marshland, and the remains of jetties far from habitation, may also indicate the locations of old Salterns. Rarely, there may be traces of the supporting buildings where the salt was actually made, its former use usually a matter of public record or simply displayed on a board outside. Most old Salterns can be seen in aerial photographs showing the rectilinear patterns of the pans, such as a group located 2 kilometres south of Holbeach St Johns in Lincolnshire, though here there is almost nothing to be seen from ground level.

There are also the 'red hills'. Although later Salterns boiled their brine in iron or (heaven help them) lead boiling pans, earlier pans were ceramic. They were large, flat and rectangular, made from clay fired at low temperature. Since they were rather fragile, they did not survive for much more than a season, and so were discarded in huge piles. The soft ceramic of discarded pans would crumble to produce the substance of these red hills. They are only called 'hills' because Salterns are always situated in low, flat areas where a 1-metre mound would be conspicuous. Red hills occur in many places, such as along the north bank of the Blackwater Estuary, just to the east of (unsurprisingly) Maldon in Essex, though sometimes they have been ploughed flat and are noticeable only from an incongruous red area of a recently worked field.

The east and south coasts of England have always been the main locations for Salterns. The slow evaporation from the open drying pans is completely negated if it rains throughout much of the production season – the season, of course, being during the warmer months of the year. The more southerly east coast of Britain is relatively dry, at around 550 millimetres of rain a year, and the warm temperatures and similarly low rainfall of the central to eastern part of the south coast were also at an advantage. Morecombe on the west coast, by contrast, receives 1,300 millimetres a year, so it was difficult, if not impossible, to produce salt on our Atlantic coasts. Southern and eastern coastlines are thus the most likely places to find Salterns.

Lincolnshire was once a major producer of salt, and, bucking the trend of disappearing into the sea, traces of Iron Age, Roman and medieval Salterns are now miles inland. The explanation for this reversal is the heroic and ultimately successful efforts to drain the fens, moving the coastline on which the Salterns were once situated. Some of these Salterns are now 20 kilometres from the sea! These are little more than faint relics recognisable only from aerial photographs and a few extant channels, but they were once major features of the landscape. There were several Salterns on the Essex coast and a large number in Hampshire, some of which are still, just about, visible. Which brings me back to Langstone Harbour.

The eastern side of Portsea Island, into which the mysterious gully ran, was, until the late seventeenth century, an area of multiple inlets, marshland and mudflats. It was reclaimed for farming and, at its southernmost part, a Saltern. It was called Great Salterns, to distinguish it from an earlier Saltern (Little Salterns) further south still at Copnor. The gully itself is now known as Great Salterns Lake, consisting of a boomerang-shaped body of water curving northward for 500 metres. The pans were situated within that curve and to the north of it, supplied with seawater from the (now) lake. The salthouse that served it still exists, on a road that now services a much less romantic out-of-town shopping complex.

Running a Saltern was very hard work, subject to the vagaries of the market, weather and taxes. Nevertheless, in the 1600s, the owner of Great Salterns, Robert Bold, did very well indeed, though not, I suspect, much of the hard graft. He became alderman in 1625 and gave Portsmouth a particularly magnificent salt cellar, now known as the Bold Salt.

A sheltered estuary or the shelter of a harbour, especially a large one with a narrow entrance such as Langstone, was necessary to protect against large waves and tidal surges. Lymington Salterns is guarded by the spit of land that ends with the Hurst Point Lighthouse and by the Isle of Wight. With Salterns, situation is (nearly) everything.

While there is a great deal not known about even the later workings of Salterns, there is a reasonable description supplied from the early nineteenth century of the workings at Lymington. The sea was admitted into large feeder ponds in which, presumably, the silty contaminants were allowed to settle. From there the water travelled via a sluice gate into a multitude of lower, partitioned pans, about 100 by 150 metres. In these, the water is kept very shallow at only about 8 centimetres and topped up when it gets low. Here, exposed to sun and wind, it becomes concentrated into a strong brine. A windmill is then employed to pump the brine into a large, raised cistern, which in turn feeds iron pans for boiling. These pans are around 2.5 metres square.

Salt-making in hot, dry climates (unlike that which pertains in Britain) does not require boiling, with evaporation alone being sufficient. In a long, dry British summer it might have been possible to dry brine sufficiently, but the considerable time it would have taken would have produced crystals the size of sugar cubes.

The boil would last for eight hours, with crystallised salt being repeatedly scooped out as soon as it formed and each pan yielding one bushel (eight gallons, equivalent to about 40 kilos) of salt per hour. The wet crystals were left in baskets to allow the bittern or bitters to drain. The bitters, if dried and crystallised, would contain a mixture of Epsom and Glauber salts, and magnesium chloride. It was necessary to remove the first two because they spoiled the clean flavour of salt (sodium chloride) and the magnesium chloride because it would deliquesce and allow the salt to become 'wet'. I still remember the less purified salt of fifty years ago, which did, invariably, become a wet or solid lump in the cellar.

Although old Salterns are relatively hard to find, and although the bulk of them are in the south and east, the existence of names like Salterns Road or Salterns Key in any coastal town or village will probably indicate that salt was once produced there. As with most of the man-made mysteries in this book, a local online search will settle the matter.

Sea Mat

[…] and roofed his mansion with silver tiles.
  Art Interchange, 1883

It is unlikely that anyone who does not take an interest in seaweed will ever notice this species sufficiently to worry about what it is. For me it is of occasional pressing concern because it grows on some of the seaweeds I like to collect to eat. Not that it is poisonous or unpleasant-tasting (other than being a little chalky); it just seems a pity to destroy organisms when it is easy not to. Of the seaweeds I collect, Carragheen, Dulse and the kelps are frequently host to this otherwise attractive little marine invertebrate.

Sea Mat, glorying in the Latin name *Membranipora membranacea*, is a Bryozoan ('moss animal') that forms dense, flat, silvery colonies on seaweeds, each colony containing thousands of individuals. In this species they each build a more or less rectangular calcium carbonate box measuring 0.5 by 0.125 millimetres, all arranged vaguely like a brick wall to maintain cohesion in the high-stress environment of the sea edge. Each little box contains a creature which collects food particles from the sea with tiny ciliated tentacles.

Piddocks

Day by day, and year by year, the piddock worked steadily on, and day by day, and year by year, the sea completed the task which the mollusc had begun.

Theodore Wood, *Sunday Magazine,* 1887

This little creature takes a belt-and-braces approach to protecting itself: as well as wearing a shell, it bores a hole in soft rock where it can not only hide but actually wedge itself immovably. On beaches where there is hard clay or soft rocks such as mudstone thousands of Piddock-holes can be seen: some occupied, some vacated by their original creator and inhabited by an empty shell or by something else completely. If you wish to know which ones contain a live Piddock, then watch as the tide goes out: if there is a Piddock in residence, it will squirt a jet of water once exposed.

There are several species of Piddock. The one I see most often is the White Piddock, *Barnea candida*, which is quite small at only 6 centimetres long. There is also the Oval Piddock, *Zirfaea crispata*, at 9 centimetres, and the Common Piddock, *Pholas dactylus*, at a substantial 14 centimetres. Piddocks are, of course, bivalves and look a little like a mussel. I ate a few once (cooked!) to see what they tasted like – just like mussels, as it happens. Apart from their covert habitat, they are easily distinguished from most other bivalves in that their shells never close completely, always gaping distinctly at the end furthest from the burrow entrance.

How do they bore their holes? The open end of Piddock shells are passingly similar to the pieces of kit one would buy from a DIY store to drill an oversize hole in a sheet of mild steel: threaded and serrated. The young Piddock firmly attaches itself to a rock with the powerful foot that extends through the opening and twists back and forth, wearing away at the soft rock

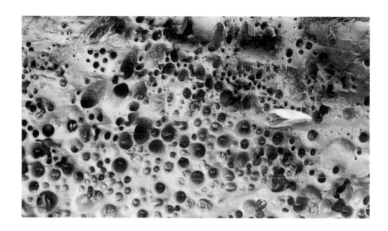

with its shell. It will continue to drill for as long as it grows – maybe five or six years. Piddocks have one other trick up their burrow, though I have never seen it: they glow in the dark. Quite why they do this I cannot imagine, and it has rather ruined my long-held principle of never eating anything that glows in the dark.

Honeycomb Worm

> *The treasures and the lair of the worm.*
> Stopford Augustus Brooke, *The History of Early English Literature: Being the History of English Poetry from its Beginnings to the Accession of King Ælfred*, Volume 1, 1892

There is almost nothing found on the seashore that does not require extensive explanation to someone – and that someone, to be fair, is often me. But I am frequently surprised at how few can recognise even the common winkle, so the various slimy, rubbery and wiggly creatures that make the seashore their home remain a mystery to most. A winkle, should you be wondering, is a green-brown sea snail, about 18 millimetres in diameter when fully grown, with a pointy top and a round opening where a little Hobbit door can be seen.

The greater number of these creatures will be found only occasionally during rock-pooling expeditions and dismissed as some sort of crustacean, worm, mollusc or jellyfish, but a few will conspicuously cover a large expanse of the shoreline and demand at least a name. One of these is the reef-making annelid worm, known as the Honeycomb Worm, *Sabellaria alveolata*. Its relatively soft, sandy reefs are made up of thousands of tubes, which in turn are made up of sand and the mucus with which each worm sticks the grains together. These 'artificial' burrows will be clustered together, each containing its maker and, in dense colonies, each tube roughly hexagonal in cross section. New burrows may be built on abandoned and collapsed burrows, so reefs sometimes reach a metre thick. On the soft and flat mudstone where I normally find this worm, they are seldom more than a few layers, but on a more robust substratum they can form huge, rounded masses. Honeycomb Worms can be found

inter-tidally and just below the low tide, built on firm substrata and with adjacent supply of sand from a sandy beach.

It is surprising how often hexagons appear in the natural world. Honeycombs are the most familiar version, but there are also snowflakes, the Giant's Causeway, the vast cloud formation at the north pole of Saturn and the rings of carbon atoms that are the basis of much organic chemistry. It is all down to efficiency – hexagons are seldom built, they just happen. The Honeycomb Worm acquires its shape in much the same way as the honeycomb itself. In the latter, bees construct cylindrical tubes out of wax. Each tube is, crucially, next to other tubes (six, in fact!). The completed tubes are warmed up by the bees, softening them and relaxing them against neighbouring tubes, naturally forming a series of hexagons. Honeycomb Worms similarly build their tubes at the same time, each one filling the available space and inevitably creating hexagons.

The Honeycomb Worm inside the tube is frankly no looker. With a large gaping mouth and more tentacles than seem strictly necessary, it is the stuff of nightmares and sci-fi movies. Tentacles, and their attached cilia, are used for feeding, and for positioning particles of sand and shells (which they coat with a glue) to the rubbery tube they form around themselves. Male animals are 30–40 millimetres long, females up to 100 millimetres. The tubes are 5–6 millimetres in diameter and at least long enough to accommodate the worm, though they can be anything up to 20 centimetres. Charmingly, each tube has a small, rounded porch.

The Honeycomb Worm lives from two to five years and breeds twice a year. The larvae feed until they are large enough to attach themselves to an existing reef. Here they will detect a chemical signature, stop swimming and settle. If they do not detect that chemical, they will attempt to settle anywhere, though most die. Considering the fact that each female may lay as many as a million eggs, it is just as well that they do. It is a fairly robust species, but still susceptible to competition, bad weather and, in the presence of man, trampling. Very much as I do when I'm walking through ancient grasslands with people and tell them not to tread on the anthills, I will always ask people to avoid trampling on Honeycomb Worms. It is possible to kill a hundred with every step!

Holes in the Sand

Life swarms with innocent monsters.
Charles Baudelaire, *Les Fleurs du mal*

My father used to take me bait-digging in Langstone Harbour, to the east of Portsmouth, looking for ragworms. Although these days out were fun enough, I still cannot look at a ragworm or

Lugworm holes and casts in the Fleet Lagoon, Dorset

think about ragworms without shuddering. Pitiable monsters, hideous to behold and nakedly vulnerable, they are miniature Grendels. My father had no qualms about threading them, alive, onto a barbed hook, but I could never contemplate imposing so grim a fate; in any case their ability to bite was discouragement enough.

Since those days I must have spent hundreds of hours looking at holes in the sand and mud on an ebbing tide, wondering what might be hiding there. Usually, wondering is all I do, but occasionally I will dig to find out. I don't particularly like doing this as, if it is a worm that I find, it will seldom survive the encounter. Frequently, however, it is a clam of some sort, and they are almost indestructible.

Identifying species according to the hole they make is a tricky task and one that is frequently impossible. But some holes are distinctive, especially when the habitat is considered as well. Another worm my father used for bait was the lugworm, which, like the ragworm, is one of many annelid worms that inhabit the seashore. It is highly conspicuous in creating two holes, one a funnel in the sand with a small hole in the bottom, the other not visibly a hole at all but a coiling pile of overcooked spaghetti made of sand. The lugworm is less terrifying than the ragworm, looking like an earthworm (also an annelid) with one end swollen and a little bristly with hairs that it uses as lungs. If you live most of your life in a burrow, it is essential that some arrangement is made for food to come in and for waste to come out. Piddocks (see p. 219), through some ingenious arrangement of plumbing, perform both functions at the same end of their short bodies. This is impossible for a worm (it's a long way from end to end), so a second hole is created for the waste. Both lugworm holes are obvious, and they are situated close to one another at just a few centimetres apart. One is that funnel in the sand or mud, complete with a distinct hole at the bottom of the funnel; the other is a 'cast', familiar to anyone who has earthworms in their lawn, an alternative comparison being wet sand squeezed out of a toothpaste tube. The burrow itself is 'U'-shaped, with the worm lying in the shape of a 'J', its bottom end at the top and its top end (head) at the bottom. It consumes seawater and sand, which descend from the funnel formed by the flow of water. It digests any organic matter that this slurry contains and the sand passes through to form the cast. There are two species of lugworm: Lugworm, *Arenicola marina*, and the Black Lugworm, *A. defodiens*. It is possible to tell them apart by examining their casts, the first producing an untidy mess, the second a neat coil. Now you know.

Ragworms have a similar arrangement, except that the holes are less distinct and their burrows sometimes more complex, adding a 'Y' to the alphabet of possible shapes. They normally live in mud and are often identifiable from the track marks they make on foraging missions from their burrows.

Before moving on from worms, there is one that lives partly in a burrow and partly in what looks like a home-made drinking straw which sticks up from the sand. It is the Sand Mason. This projects about 3 centimetres from the sand and is covered with sand particles for protection.

With the exception of a truly appalling dish of annelid worms apparently enjoyed in Vietnam where they are considered a delicacy, marine worms are not eaten, well, anywhere. Clams, however, are extremely good to eat, which is the reason I happen to know a bit about how to find them. All clams are filter-feeders, sucking in seawater, filtering out phytoplankton and squirting out what is left. Those that live in sand or mud have siphons which allow them to hide in the seabed while sucking and blowing at leisure. The holes these make are sometimes visible and distinctive from the variety of clam size, shapes and feeding habits.

Cockles can often be found just lying around on the seabed or shore, but most will be dug in. To imagine *how* they are dug in, hold your hands together in prayer and in front of your face, then open them like a book. This gives you an idea of their orientation respective to the hinge and the position of the siphons, one of which will be more or less in line with your third finger, the other in line with your middle finger. The two siphons are very short, so the cockle edge needs to be just at the surface in order to feed. With the siphons also being very close together, the hole in the sand (and sometimes mud) is small and untidy, and both siphons and shells are sometimes visible.

The Soft-Shell Clam, *Mya arenaria*, is a large grey clam of muddy estuarine habitats. Quite unlike the Cockle, it digs in deep, as far as 30 centimetres, and has a correspondingly long and muscular siphon. This leaves a hole in the mud which is oval and large, at 5 millimetres in diameter. The Grooved Carpet Shell, *Tapes decussatus*, is similar in appearance to the Cockle, if a little more elegant, and slightly larger. Its two siphons are separated, resulting in two holes about 2 centimetres apart. It is a great joy to plunge your hand into the sand and come up with one of these tasty treasures.

Finally, in this far from complete review of Clam siphon holes, there are the Razor Clams. There are five or six species in British waters, but they all create a very distinct hole as a giveaway to their position. Given the shape of the clam and vertical orientation, the two siphons are, inevitably, close together. The hole is a distinct 'keyhole' or figure-of-eight in the sand. Unlikely as it may seem with something that spends its entire life in salt water, a sprinkle of salt is enough to irritate them sufficiently to rise like Aphrodite.

Further Reading

Tom Cope and Alan Gray, *Grasses of the British Isles,*
BSBI Handbooks (London, 2009)
The standard treatment, but it is not an easy subject and if you want to get to grips with it, buy more books and be prepared to put in a great deal of work.

Martin B. Ellis and J. Pamela Ellis, *Microfungi on Land Plants: An Identification Handbook* **(Slough, 1997)**
I was at the book launch party for this remarkable book, back in 1997 – a proud moment for me. As with plant gall identification, it is possible to look up the host and decide between the relatively few possibilities.

Simon Harrap, *Harrap's Wild Flowers* **(London, 2013)**
Provides excellent photographs all of the plants you are likely to find and jargon-free descriptions that are helpful for identification.

Peter J. Hayward and John S. Ryland (eds.), *Handbook of the Marine Fauna of North-West Europe* **(Oxford, 2017)**
From a 2-millimetre marine snail to a metre-long fish, nearly all of the animals found on or near the beach are in this book. A comprehensive guide that you would not wish to drop on your toe.

Jessica Holm and Guy Troughton, *Squirrels*, British Natural History Series (London, 1987)
Everything you want to know about squirrels, except how to cook them.

Bruce Ing, *The Myxomycetes of Britain and Ireland: An Identification Handbook* (Slough, 1999)
Just about the only book that covers the domestic species, and a must-have for budding myxomycophiles.

Peter Marren, *Mushrooms: The Natural and Human World of British Fungi* (London, 2019)
A book from the much-respected New Naturalist series, this is an excellent and readable overview of the biology and ecology of fungi.

C. S. Orwin and C. S. Orwin, *The Open Fields* (Oxford, 1954)
An old book, but worth a read to understand this thousand-year agricultural system.

Oliver Rackham, *History of the Countryside* (London, 1986)
Rackham provides a very readable overview of the British countryside, solving many puzzles along the way.

Margaret Redfern and Peter Shirley, *British Plant Galls*, p/b edn (Dorking, 2011)
The standard work on British galls and very easy to use, as you simply look up the plant on which the gall is growing and go through the fairly small number of possibilities.

John S. Rodwell (ed.), *British Plant Communities*, 5 vols. (Cambridge, 1991–2000)
This remarkable series categorises the various habitat types (communities) found in grasslands and woodlands as well as wetlands, aquatic and maritime habitats. It is used by professional and serious amateur naturalists to categorise any natural or semi-natural area of land in Britain.

Francis Rose, *The Wild Flower Key: How to Identify Wild Plants, Trees and Shrubs in Britain and Ireland*, rev. edn (London, 2006)
While almost indispensable for the intermediate botanical student, being a *key* where a series of questions is posed, which, when answered accurately, will (eventually) provide an identification. It is not an easy read. The drawings are helpful and include definitive line drawings of things like seed capsules.

Clive A. Stace, *New Flora of the British Isles*, 3rd edn. (Cambridge, 2010)
Comprehensive, dense and not shy in its use of jargon, this is the gold standard of British botany and the book that nearly every serious botanist will own. Don't expect too many drawings or photographs as they're fairly sparse, so it is not a book than can be flicked through until you see something that 'looks a bit like' the plant you have in your hand.

Paul Sterry and Andrew Cleave, *Collins Complete Guide to British Coastal Wildlife* (London, 2012)
Plants, lichens, seaweeds, worms, marine molluscs, jellyfish, sea slugs, crustaceans and littoral fish are all here in this essential book for any trip to the seaside.

Bob Watson, *Trees: Their Use, Management, Cultivation and Biology – A Comprehensive Guide* **(Ramsbury, 2006)**
If you are interested in trees and how they grow and live, it is all here. Nicely illustrated, this is a much-used favourite on my bookshelf.

Tom Williamson, *The Archaeology of Rabbit Warrens* **(Princes Risborough, 2006)**
A relatively short and readable book that goes into considerable depth on the subject of pillow mounds.

Michael Wingate, *Small-Scale Limeburning: A Practical Introduction* **(London, 1985)**
This book goes into much more detail than my brief overview.

Index

Acknowledgements

I needed some serious help with many of the subjects in this book and have received it unstintingly from friends who are all too used to getting a call from me. But first of all, profound thanks go to my two tireless researchers, William Evans and Lily Wright, who did much of the heavy lifting, though I rather fancy they enjoyed it.

Many thanks to the extraordinary natural-history polymath, Bryan Edwards, who has, yet again, provided me with much invaluable information and put me right about a few things without a hint of scorn, to Eddie Bailey for his encouragement and for sharing his vast knowledge of geology, to Neville Kilkenny and Roy Watling for the desperately needed help they gave me with the trickier bits of mycology and to Richard Pearson for his inspired suggestions.

My thanks are due to all those from Profile Books who have worked on the book. Patience in an editor is a rare and precious thing, so thank you Louisa Dunnigan, for this and for everything – especially for allowing me more than my normal quota of jokes. I am extremely grateful to Matthew Taylor for his cheerful and indulgent precision as copy-editor. Thanks also to Lottie Fyfe and James Alexander for the excellent work they have done on the layouts despite my constant interventions, and to designer Peter Dyer and artist Clare Curtis for producing such a truly delightful cover. In the confident anticipation that comes only from happy experience, I would like to express my gratitude to Ruth Killick, my publicist for the book.

Finally, as ever, thanks are duly given to my agent, Gordon Wise of Curtis Brown, for his continuing and ever-enthusiastic support.